Christian Bilik

Lightweight Structural Methods

Christian Bilik

Lightweight Structural Methods
for Increasing the Buckling Load and Fundamental Frequency of Thin-walled Structures

Südwestdeutscher Verlag für Hochschulschriften

Impressum/Imprint (nur für Deutschland/only for Germany)
Bibliografische Information der Deutschen Nationalbibliothek: Die Deutsche Nationalbibliothek verzeichnet diese Publikation in der Deutschen Nationalbibliografie; detaillierte bibliografische Daten sind im Internet über http://dnb.d-nb.de abrufbar.
Alle in diesem Buch genannten Marken und Produktnamen unterliegen warenzeichen-, marken- oder patentrechtlichem Schutz bzw. sind Warenzeichen oder eingetragene Warenzeichen der jeweiligen Inhaber. Die Wiedergabe von Marken, Produktnamen, Gebrauchsnamen, Handelsnamen, Warenbezeichnungen u.s.w. in diesem Werk berechtigt auch ohne besondere Kennzeichnung nicht zu der Annahme, dass solche Namen im Sinne der Warenzeichen- und Markenschutzgesetzgebung als frei zu betrachten wären und daher von jedermann benutzt werden dürften.

Coverbild: www.ingimage.com

Verlag: Südwestdeutscher Verlag für Hochschulschriften GmbH & Co. KG
Heinrich-Böcking-Str. 6-8, 66121 Saarbrücken, Deutschland
Telefon +49 681 37 20 271-1, Telefax +49 681 37 20 271-0
Email: info@svh-verlag.de

Approved by: Wien, TU, Diss., 2011

Herstellung in Deutschland:
Schaltungsdienst Lange o.H.G., Berlin
Books on Demand GmbH, Norderstedt
Reha GmbH, Saarbrücken
Amazon Distribution GmbH, Leipzig
ISBN: 978-3-8381-2978-5

Imprint (only for USA, GB)
Bibliographic information published by the Deutsche Nationalbibliothek: The Deutsche Nationalbibliothek lists this publication in the Deutsche Nationalbibliografie; detailed bibliographic data are available in the Internet at http://dnb.d-nb.de.
Any brand names and product names mentioned in this book are subject to trademark, brand or patent protection and are trademarks or registered trademarks of their respective holders. The use of brand names, product names, common names, trade names, product descriptions etc. even without a particular marking in this works is in no way to be construed to mean that such names may be regarded as unrestricted in respect of trademark and brand protection legislation and could thus be used by anyone.

Cover image: www.ingimage.com

Publisher: Südwestdeutscher Verlag für Hochschulschriften GmbH & Co. KG
Heinrich-Böcking-Str. 6-8, 66121 Saarbrücken, Germany
Phone +49 681 37 20 271-1, Fax +49 681 37 20 271-0
Email: info@svh-verlag.de

Printed in the U.S.A.
Printed in the U.K. by (see last page)
ISBN: 978-3-8381-2978-5

Copyright © 2012 by the author and Südwestdeutscher Verlag für Hochschulschriften GmbH & Co. KG and licensors
All rights reserved. Saarbrücken 2012

Abstract

Stability loss of a structure demonstrates a mode of failure which preferentially occurs for slender or thin-walled structures as typical in the lightweight design. For increasing critical loads or enhancing safety margins for given loads, constructive measures, especially the usage of stiffeners, are often required. Similar holds for the eigenfrequency representing a dynamic characteristic of a structure. Under usage structural eigenfrequencies of a structure should not be excited in order to avoid resonance phenomena. Also, in this case, design solutions often comprise the application of stiffeners to increase the eigenfrequencies and shift them to higher frequency-domains, which are not being excited by operating states. Typically an installation of stiffeners results in a mass increase as well as a geometry change of the concerned structure.

Aim of the thesis at hand is the investigation and evaluation of different possibilities of structural modifications to enhance the buckling stability. Further goal is a raise of the eigenfrequency in terms of increasing a structure's fundamental frequency. Both of these aims should be achieved without mass increase and with either no or very small changes in geometry. Finally the manufacturing process roll forming is investigated in terms of stability of the resulting product and the effects of process-related residual stresses present in the final product.

It can be shown that residual stresses, contained in a part or structure, can act as stiffeners. Structures containing advantageous residual stresses (in present case laser irradiation is used as thermal treatment to introduce residual stresses) are stiffened according to their buckling behaviour and in terms of an enhancement of their eigenfrequency without the necessity of adding additional mass. Different kinds of laser treatment result in different residual stress distributions within the affected structure. Deformations during and after the laser treatment should be kept as small as possible. Further investigations are comparisons with experimentally found results.

An introduction of beads, representing a classical, geometry-changing stiffening method for thin-walled structures is also investigated within the current work. Presently geometries in that case are predominantly gained by trial and error principle or found in literature. A bead laying algorithm for finding suitable geometry pathes for the beads is described and applied to different structures.

During the roll forming process, representing a continuous forming process of thin metall strips, the undesired occurence of instabilities during the forming process as well as in the deformed semi-finished product is possible. Additionally, residual stresses within the finished product can result in significant changes in its stability- and eigenfrequency behaviour. Both is investigated in the present work under usage of numerical methods.

Keywords: buckling, laser treatment, residual stresses, roll forming, beads, natural vibrations

Acknowledgment

The thesis at hand was conducted at the Institute of Lightweight Design and Structural Biomechanics, Vienna University of Technology under supervision of O.Univ.Prof. Dipl.-Ing. Dr. Franz G. Rammerstorfer.

The financial support by the Austrian Federal Government (in particular from the BMVIT and the BMWFJ) and the Styrian Provincial Government, represented by FFG and by SFG, within the research activities of the K2 Competence Centre MPPE, operated by MCL Leoben in the framework of the Austrian COMET Programme, is gratefully acknowledged.

Contents

1 Introduction — 9
   1.1 Aim and Outline of the Thesis 9

2 Linear Structural Stability and Linear Structural Eigenfrequency — 11
   2.1 Introduction 11
   2.2 Analytical Approach 12
      2.2.1 Buckling of Bars 12
      2.2.2 Buckling of Plates 14
   2.3 Evaluation Method for Buckling Experiments 15
   2.4 Eigenfrequency Calculation and Measurement 15

3 Residual Stresses — 18
   3.1 Introduction 18
   3.2 Formation of Residual Stresses by Plastic Deformation 19
   3.3 Formation of Residual Stresses under Heat Treatment 20
   3.4 Effect of Residual Stresses on the Stability- and Eigenfrequency Behaviour of Structures 24
      3.4.1 Stability 24
      3.4.2 Eigenfrequency 25
   3.5 Experimental Investigation of Residual Stresses 25

4 Laser Treatment of Plate-Structures — 27
   4.1 Introduction 27
   4.2 Modelling of Laser Heat Treatment of Plate-Structures by the Finite Element Method 28
      4.2.1 Specifying the Model-Type 28
      4.2.2 Material Data 29
      4.2.3 Modelling of the Laser Heat Input 32
      4.2.4 Development of Residual Stresses, FEM Investigations of Laser Heat Treatment with a single Laser Point 33
   4.3 Analysis of Laser Treated Plates 64
      4.3.1 Continuously Laser Treated Plates 68
      4.3.2 Pointwise Laser Treated Plates Type 25P 76
   4.4 Evaluation of Laser Treatments with Respect to Buckling Resistance — 81

		4.4.1	Evaluation according to Buckling Load	82
		4.4.2	Evaluation according to the Fundamental Frequency	93
	4.5	Analyses of Stiffened Plates with Laser Treatment	103
		4.5.1	Case 1, untreated reference plates	109
		4.5.2	Case 2, unstiffened plates under laser treatment	109
		4.5.3	Case 3, laser treatment of unstiffened plates, beam stiffeners are mounted after completed laser treatment	118
		4.5.4	Case 4, laser treatment of unstiffened plates, out of plane deformations (direction z) on the edges are reset to zero, beam stiffeners are mounted thereafter	123
		4.5.5	Case 5, laser treatment of already stiffenend plates	126
		4.5.6	Conclusion for laser treatment of prestiffened plate-structures	126
	4.6	Comparison with Experimental Results of Laser Treated Plates . . .	129
		4.6.1	Experimental Laser Treatment of Plates	129
		4.6.2	Test Equipment	130
		4.6.3	Comparing Simulation and Experiment	131
	4.7	Discussion of obtained Results	132

5 Bead Laying under Optimisation Aspects   136

	5.1	Introduction of Beads into Structures	136
		5.1.1 Manufacturing Principles	137
	5.2	The Bead Laying Algorithm	138
	5.3	Application of the Bead Laying Algorithm	146
		5.3.1 Application to a Flat Structure - The Plate	146
		5.3.2 Application to a Profile Structure - The U-shaped Profile . . .	150
	5.4	Conclusions	150

6 Roll Forming   157

	6.1	Introduction	157
	6.2	Literature Review	158
		6.2.1 Literature - Experimental Data from the Roll Forming Process	158
		6.2.2 Literature - Analytical Methods for Modelling	158
		6.2.3 Literature - Modelling based on the Finite Element Method	159
		6.2.4 Literature - Residual Stresses in Formed Products	161
		6.2.5 Summary of Literature Overview	161

Contents

6.3	Stability Problems in Connection with the Roll Forming of Profiles	162
	6.3.1 FEM Model representing the Roll Forming Process	163
	6.3.2 Results of the FEM Model for the Roll Forming Process Simulation	163
6.4	Influence of Residual Stresses due to Roll Forming on the Stability of the Rollformed Product	172
	6.4.1 Roll Forming Mill for an U-shaped Profile	172
	6.4.2 Cutting of the U-shaped Profile	176
	6.4.3 Comparison with 'perfect' Residual Stress free Profile	178
6.5	Summary and Conclusion	184

Contents

Chapter 1

Introduction

1.1 Aim and Outline of the Thesis

The thesis at hand results from a scientific project within the COMET K2 Centre MPPE (operated by the MCL - Materials Center Leoben, Forschung GMBH) with the Lehrstuhl für Umformtechnik (LUT), Department Product Engineering at the Montanuniversitaet Leoben (MUL) and the Institute of Lightweight Design and Structural Biomechanics (ILSB), Faculty of Mechanical and Industrial Engineering at the Vienna University of Technology (TUW) being involved as scientific partners. Project title of this mentioned project A3.9 is 'Metal Forming Concepts for Manufacturing of Light Weight Structures'.

One of the main scientific objectives of the project is the investigation of possible methods to increase the structural stability of thin shell structures. The systematic introduction of residual stresses into such a structure represents one possible method. In that case, deeper investigations of the key-influences of residual stresses are conducted. Representing a further stiffening method the introduction of bead patterns into a structure is investigated. Another objective of the project concerns the manufacturing process roll forming. Also here, investigative emphases are layed on the occurence of stability problems during the process itself as well as on resulting residual stresses within the roll formed products and their influence onto the product's stability behaviour.

Starting with an introduction, a short overview about the loss of structural integrity with a main emphasis on plate-buckling and the structural eigenfrequency behaviour of plates is given in Chapter 2. Chapter 3 deals in general with residual stresses and

Chapter 1 Introduction

in particular with their development, distribution and effect on structural stability behaviour of plates. In Chapter 4 the introduction of residual stresses into plates by means of laser treatment is being investigated. Stiffening effects of beads for increasing buckling loads, in terms of an incremental algorithm based enhancement, is topic of Chapter 5. In Chapter 6 the manufacturing process 'roll forming' is analysed in regard to the stability behaviour of the resulting products which contain residual stresses.

Chapter 2

Linear Structural Stability and Linear Structural Eigenfrequency

2.1 Introduction

Today many industrial products and components are revised and improved to reduce their weight because of economic and environmental protection reasons. In terms of optimisation one would speak of optimising a structure with the objective function being the mass. In classical mechanical engineering design, limiting conditions to constructions are often certain mechanical stress levels or deformation limits which must not be surpassed. Weight of the structure was of course always an issue but maybe not the main one and was rather neglected in favour of having quite high safety margins. With the availability of modern high strength materials in many cases, a significant reduction of thickness and therefore saving in mass is possible. Through the resulting change in constructions, limiting conditions can change and other problems may arise.

Lightweight designs often tend to fail by reason of stability loss rather than by exceeding certain stress levels. Another important issue is the sudden unexpected occurence of such a stability loss which may result in unwanted or dramatic conditions for the structure affected. One speaks of either stable or unstable post buckling behaviour meaning the structural integrity is still present after surpassing the stability limit or it fails dramatically. Plate structures are representative for exhibiting a stable elastic post buckling behaviour whereas cylindrical shell structures show an unstable post buckling behaviour. Nevertheless a stable post buckling behaviour can also lead to unwanted system states and therefore has to be anticipated. Speaking of

Chapter 2 Linear Structural Stability and Linear Structural Eigenfrequency

stability loss of the equilibrium state of a perfect structure leads from mathematical point of view to an eigenvalue problem in the case of bifurcation buckling. Solving this eigenvalue problem leads to eigenvalues, representing the point of branching off from the stable path (in a load displacement diagram), and eigenvectors, representing the corresponding mode-shape of the problem. However in nature as well as in applied technologies no perfect structures can be found. Thus, it is often important to take possible imperfections within the structure into account. A well suited method to characterise a structure's behaviour, which also might be rather imperfect, is to prepare a load-displacement graph. Within a load-displacement graph a load parameter, representing the intensity of externally applied loads, is displayed over the displacement of a suitable point. In Fig. 2.1 a typical load displacement behaviour of a perfect compared to an imperfect plate-structure (with increasing degree of imperfection) is shown. As mentioned above, plates exhibit a stable post buckling behaviour. Thus, a complete loss of the structure's load carrying capacity does not take place, as long as the structure behaves elastically.

2.2 Analytical Approach

2.2.1 Buckling of Bars

Euler-buckling for beams can be considered as classical example for a stability problem. As shown in [47], a consideration of the differential equation Eq. 2.1 with load parameter $\alpha^2 := \frac{P}{EJ}$ (P representing the load in axial direction of the bar) leads for bars with a constant cross section to Eq. 2.2 with the solution given in Eq. 2.3.

$$EJw'''' + Pw'' = 0 \qquad (2.1)$$

$$w'' + \alpha^2 w = 0 \qquad (2.2)$$

$$w = A cos(\alpha x) + B sin(\alpha x) \qquad (2.3)$$

Further considerations of boundary conditions lead to the nontrivial solutions for α as given in Eq. 2.4.

$$\alpha_n^* = \frac{n\pi}{l}, n = 1, 2, ... \qquad (2.4)$$

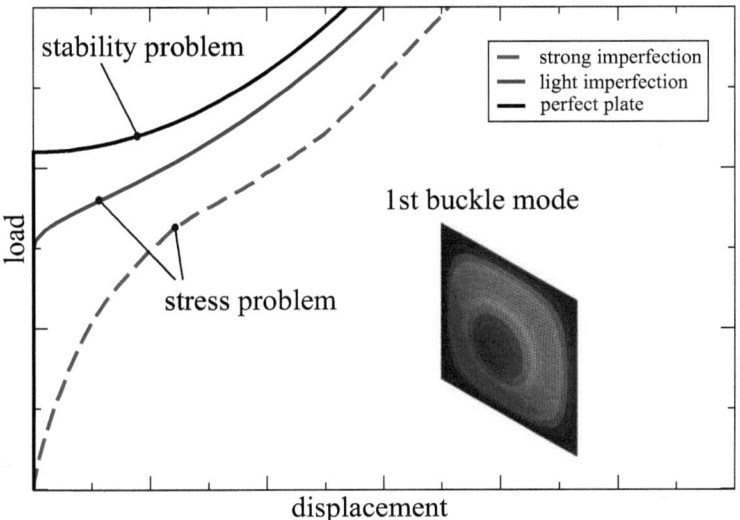

Figure 2.1: Comparison of load-displacement behaviour of different (perfect an imperfect) plate-strutures.

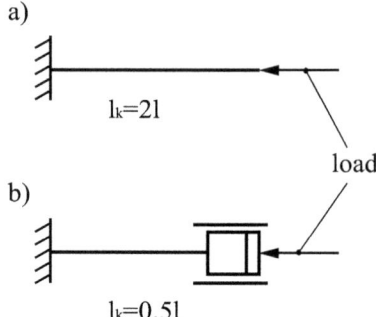

Figure 2.2: Example for the dependency of the effective length l_K on the boundary conditions of a bar. Case a) one side free, case b) one side guided support.

Finally the Euler-buckling formula for homogeneous bars, possessing a constant cross section, is given in Eq. 2.5 with E being the Young's modulus, J the moment of inertia an l_K representing the effective length of the bar.

This effective length l_K has a quadratic influence on the critical load P_K^*, see Eq. 2.5, and shows a strong dependency on the respective boundary conditions. For example, l_K is reduced from two times the bar length to 0.5 times the bar length if the loaded bar end (the other side is clamped) is not free but has a guided support, see the sketched situation shown in Fig. 2.2.

$$P_K^* = \frac{\pi^2 E J}{l_K^2} \tag{2.5}$$

2.2.2 Buckling of Plates

Also taken from [47] the differential equation prescribing buckling of rectangular plates is given in Eq. 2.6.

$$K\Delta\Delta w + N_{xx}\frac{\partial^2 w}{\partial x^2} + N_{yy}\frac{\partial^2 w}{\partial y^2} + 2N_{xy}\frac{\partial^2 w}{\partial x \partial y} = 0 \tag{2.6}$$

Chapter 2 Linear Structural Stability and Linear Structural Eigenfrequency

Without going into details the solution of mentioned differential equation follows as shown in Eq. 2.7 with k being the buckling factor.

$$\sigma^* = kE\left(\frac{t}{b}\right)^2 \tag{2.7}$$

Buckling factors are available for many kinds of plate boundary and loading conditions in diagrams, further information is given in [47], too. These formulas are mentioned here because they represent the method of analytical treatment used for all perfect plate-structures in subsequent investigations of this thesis, representing also the reference state.

2.3 Evaluation Method for Buckling Experiments

For comparing results obtained from either analytical calculations (if those are available, e.g., for plate- and cylindrical shell structures) or FEM-simulations (e.g., buckling analysis) with experimental data, one could use Southwell's method. Originally, Southwell's method was developed for evaluating experimentally obtained load-displacement pathes of (naturally imperfect) buckling structures with respect to the critical load of the structure without imperfections, see eg. [54].

In Fig. 2.3 a sketch of a Southwell diagram is depicted, like it is presented in [3]. On the horizontal axis a defined displacement a is applied, for example an out-of-plane displacement of a plate-structure's midpoint resulting from a load P. The fraction 'displacement over load' $f(a) = \frac{a}{P}$ is applied to the vertical axis. With values a and P obtained, e.g., from experimental investigations, the Southwell diagram can be sketched by using a linear regression for the data points. Under application of Eq. 2.8 one obtains the critical load P_K^* of the structure. Further result is a_0, the predeformation of the structure, see Fig. 2.3.

$$\frac{df(a)}{da} = \frac{1}{P_K^*} \tag{2.8}$$

2.4 Eigenfrequency Calculation and Measurement

Eigenfrequency represents a basic parameter of a structure and is part of its dynamic behaviour. This parameter is important to know because under vibration excita-

Chapter 2 Linear Structural Stability and Linear Structural Eigenfrequency

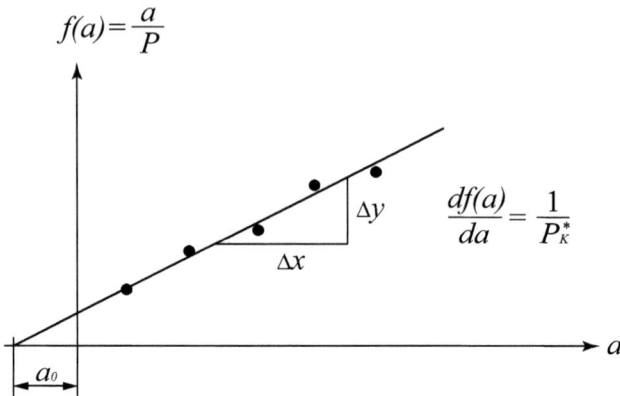

Figure 2.3: Schematic example of a Southwell diagram. Input: a displacement under load, P load applied on structure. Output: P_K^* structure's critical load, a_0 predeformation of the structure.

tion no resonance phenomenon should occur for a part (or construction). Thus, the respective eigenfrequencies should be out of the range of service excitation. If the first eigenfrequency is already above the range of service excitation all other harmonic frequencies are higher too. The first or fundamental eigenfrequency has a close linkage to the stability behaviour of a structure. For conservative systems load situations leading to a vanishing fundamental eigenfrequency correspond to critical loads in terms of buckling. Therefore this can be used in an indirect tracking of a stability loss (in terms of buckling), e.g., in an accompanying eigenfrequency measurement during a load increase (this concept is used in Chapter 4.4.2). Detailed correlations between an increase in the fundamental frequency (representing the first eigenfrequency) and an increase of the buckling load can be found in [46].

Analytical solutions to obtain bending-eigenfrequencies of a residual stress-free and unloaded plate under usage of Kirchhoff's plate theory are for example given in [40] and are presented in Eq. 2.9.

$$\omega_{m,n} = \pi^2 \left[\left(\frac{m}{a}\right)^2 + \left(\frac{n}{b}\right)^2 \right] \sqrt{\frac{D}{\rho h}} \qquad (2.9)$$

Chapter 2 Linear Structural Stability and Linear Structural Eigenfrequency

In this equation m and n, respectively, are the half-wave numbers of the occuring vibration mode in x- and in y- direction. Values a and b represent the plate's edges length. Further parameters are ρ, representing the material's mass density and h being the thickness of the plate-structure. The bending stiffness is given by D which results from the Young's modulus E and the Poisson's ratio ν, see Eq. 2.10.

$$D = \frac{Eh^3}{12(1-\nu^2)}. \tag{2.10}$$

These analytical equations are used to check FEM results for untreated square plates simply supported at all edges, presented in the respective forthcoming Chapters.

Not only FEM simulations are performed to calculate eigenfrequencies but also experimental investigations are carried out by the LUT in Leoben. In this case the object of investigation is excited by an impuls caused by an impact of a small hammer. The time development of accelerations in the excited structure is then measured with an piezo acceleration sensor. An evaluation of the received acceleration signal is done via an FFT (Fast Fourier Transformation) analysis in the software programme Matlab[1]. Resulting diagrams (in the frequency domain) show peaks at distinct frequencies, the eigenfrequencies. A good overview of frequency analysis can be found, e.g., in [19] or in [29].

---

[1]The MathWorks, Inc., 3 Apple Hill Drive Natick, MA 01760-2098, USA

Chapter 3

Residual Stresses

3.1 Introduction

Residual stresses are found in many components of technical interest. Especially heat treated parts which are subjected to non-uniform cooling down (e.g. cast parts) often exhibit residual stresses. Residual stresses in materials are generally distinguished into different types. According to [45] (which gives a good overview about the development and possible measurement of residual stresses) residual stresses of type 1 are a kind of material tensioning of macroscopic nature and are constant within the [mm] length scale. Further residual stresses ('higher' types 2 to 4) are, according to [45], within the cristal-domain or even downscaled to the atomistic scale. Investigations carried out in the objective thesis only deal with residual stresses of type 1.

Literature about the development and the evaluation as well as the calculation of residual stresses is for example given in [39] or [48]. For the development of residual stresses during laser-welding (which is related to the treatment investigated in Chapter 4) detailed information is given, e.g., in [44]. Furthermore [22], [42] and [49] give comparisons between experimentally determined residual stresses and results obtained from simulations and calculations. In [53] influences on the accuracy of experimental residual stress investigations are given. A comparison of different methods to perform residual stress measurements, with the emphasis on welding seams, is given, e.g., in [8].

3.2 Formation of Residual Stresses by Plastic Deformation

One of the main possibilities how residual stresses can be introduced into a structure is represented by plastic deformations. Hereby the material is deformed beyond it's elastic range, and plastic deformations occur. In the simplest case after unloading the stress state is rearranged and reduced by its linear component. This may also result in a stress state which can be rather uneven. In [47] an example for a beam (consisting of elastoplastic material) under bending load is given.

As an example a FEM model for a T-joint is shown in the following. Here a T-joint connection consisting of two hollow profiles with square cross section is deformed according to a prescribed displacement imposed on the vertical profile's top. In Fig. 3.1 the loading as well as the boundary conditions, which are present in the model, are shown. Hereby points A, B and C have locked boundary conditions in the directions which are indicated in the brackets. The loading of this structure is a prescribed displacement of point C into direction z. As results of the analysis von Mises stress states are shown for three conditions, see Fig. 3.2. Condition one represents the unloaded state. Conditon two shows the stress state after the full prescribed displacement of point C is applied and the point is still held in this position. Condition three shows the state of the residual stresses after release of the precribed displacement of point C (resulting in an elastic springback).

The second FEM model given here as an example deals with embossing. Embossing is a special metal forming procedure, which is for example used to produce coins or to introduce different patterns into plate-structures. Figure 3.3 shows results obtained from the simulation. In this depicted example, circular marks are embossed (by using a combination of a convex punch and a concave counter support) into a plate-structure. One can see that after the deformation process residual stresses remain in the vicinity of each deformation mark.

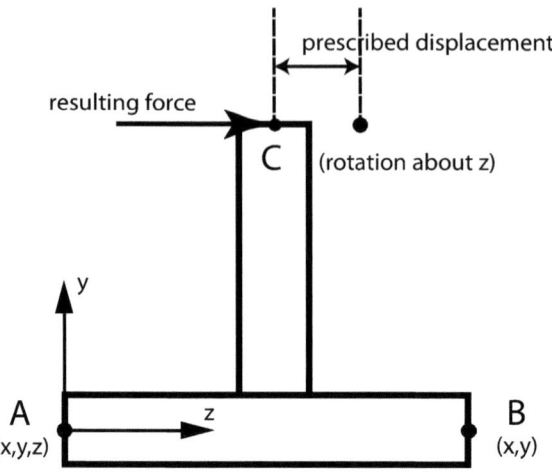

Figure 3.1: Boundary and loading conditions present at the T-joint FEM model.

3.3 Formation of Residual Stresses under Heat Treatment

As already mentioned before, heat treatment can possibly result in the development of residual stresses in a structure if heat induced stresses exceed yield limit. In Fig. 3.4 a snapshot of a FEM simulation, which shows a continuous laser heat treatment along a straight track (more details are presented in forthcoming Chapter 4), is depicted. Hereby in an uneven, time dependent temperature field yield stresses are locally exceeded by the equivalent stress corresponding to the thermal stresses. This is leading to plastic deformations, which in combination with the non-uniform cooling down, lead to residual stresses finally existing in the structure.

Chapter 3 Residual Stresses

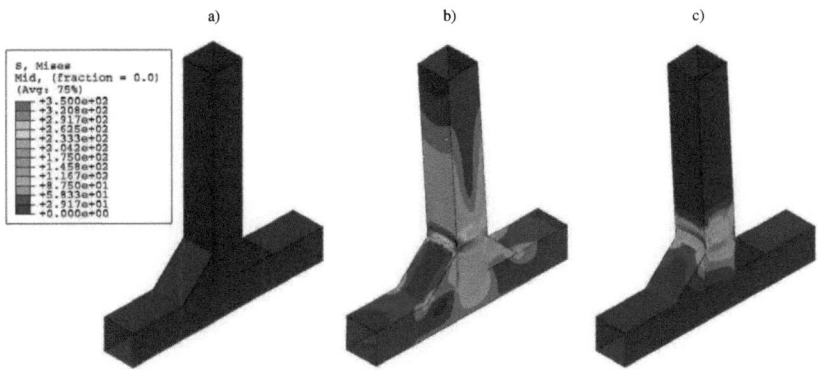

Figure 3.2: Deformation procedure on an T-joint connection. Picture a): Unloaded state. Picture b): Stress state after full prescribed displacement, point C still held in position. Picture c): Residual stress state after complete relief of the prescribed displacement. (Valid for all three pictures: von Mises stresses in MPa, each in the structure's mid-plane)

Chapter 3 Residual Stresses

Figure 3.3: Residual stress state (von Mises stresses in MPa) after embossing at three positions in a plate-structure. Stress state depicted in the structure's mid-plane.

Chapter 3 Residual Stresses

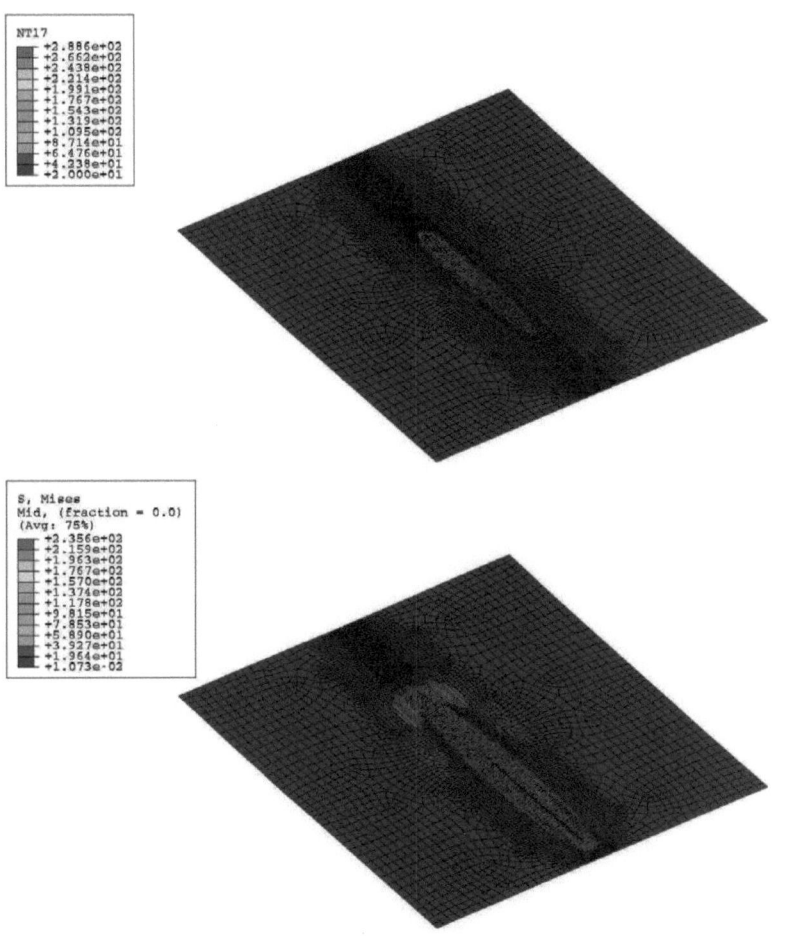

Figure 3.4: Top picture: Temperature field (in °C at the plate-structure's top surface) during continuous laser irradiation along a straight laser track. Bottom picture: Corresponding von Mises stress field (in MPa) occuring at the same time.

3.4 Effect of Residual Stresses on the Stability- and Eigenfrequency Behaviour of Structures

Enhancing the stability behaviour and raising the fundamental frequency of a structure are part of the main topics of the thesis at hand, as already stated before. As for example shown in [46], appropriate residual stress distributions within a structure can possibly achieve both desired improvements. Evaluating such residual stress distributions in a structure is in the current case done numerically under the usage of FEM models. Mainly the commercial FEM software ABAQUS (e.g., in the version [2]) is applied, allowing a broad range of analyses. Among these, perturbation type analyses for both buckling investigation and also eigenfrequency analyses represent capable methods for the given tasks.

Perturbation means in this case the application of small disturbances onto a system's state. These disturbances are applied on a linearised problem (linearised at a given value). For both the buckling analysis and the eigenfrequency analysis it leads to eigenvalue problems from a mathematical point of view.

3.4.1 Stability

With the notation used in the manual of the FEM programme ABAQUS, see [2], the eigenvalue problem is given in the form of Eq. 3.1 (i refers to the ith buckling mode, the eigenvector $\vec{v}_i$ corresponds therefore to the ith eigenvalue λ_i) and $\boldsymbol{K}_0$ represents the stiffness matrix of the structure in the base state, i.e., the equilibrium state which is considered with respect to its stability. This base state can include preceding deformations imposed onto the structure as well as preloads P which then result in a critical load $P^* = P + \lambda_i Q$. In this equation Q represents the reference load which is multiplied by the eigenvalue λ and superimposed with the preload P finally leading to the critical load P^*. In this kind of consideration, residual stresses are included in the matrix $\boldsymbol{K}_0$. Furthermore, matrix $\boldsymbol{K}_\Delta$ represents the change in the tangent stiffness matrix due to the applied reference load Q.

$$(\boldsymbol{K}_0 + \lambda_i \boldsymbol{K}_\Delta)\vec{v}_i = 0 \qquad (3.1)$$

A more detailed view on the influence of residual stresses on the tangential stiffness matrix of a structure can be found in [5], see Eq. 3.2, presented in a simplified way.

$$\boldsymbol{K}_0 = \boldsymbol{K}_e + \boldsymbol{K}_g \tag{3.2}$$

The tangential stiffness matrix $\boldsymbol{K}_0$ consists of $\boldsymbol{K}_e$ representing that part of the tangential stiffness matrix which depends explicitly on the geometry and the current material state as well as on the deformations appearing in the base state. Further part of $\boldsymbol{K}_0$ is $\boldsymbol{K}_g$ being the geometric stiffness matrix (depending on geometry and explicitly on the current stress-state which contains the residual stresses in addition to the stresses caused by the external load P).

3.4.2 Eigenfrequency

For the eigenfrequency of small vibrations, i.e., small amplitudes, at a certain load and displacement state, the eigenvalue problem given in Eq. 3.3 holds. M is the mass matrix and ω_i is the angular frequency according to the ith vibration mode. Finally, for the eigenfrequency, in terms of cycles per time unit, the relationship $f_i = \frac{\omega_i}{2\pi}$ is valid.

$$(\boldsymbol{K}_0 - \omega_i^2 \boldsymbol{M})\vec{\phi}_i = \boldsymbol{0} \tag{3.3}$$

Also in this case the effect of residual stresses is included in $\boldsymbol{K}_g$, see Eq. 3.2.

3.5 Experimental Investigation of Residual Stresses

Destructive as well as non-destructive methods, for measuring residual stresses, were applied in order to validate results obtained by simulations.

The borehole method represents the destructive method which was used. An assembly of a high speed milling machine on the laser treated plates (Chapter 4), shown in Fig. 3.5, is used to perform the borehole measuring method. Quite good agreement with values obtained from simulations were found, as also stated in forthcoming Chapter 4. Detailed information about this comparison is given in [23].

X-ray diffractometry, representing a non destructive method for measuring residual stresses, is also performed in [41], the experimental equipment is shown in Fig. 3.6.

Chapter 3 Residual Stresses

Figure 3.5: High speed milling machine for applying the borehole method to measure residual stresses. (courtesy of G. Figala, LUT, Montanuniversitaet Leoben)

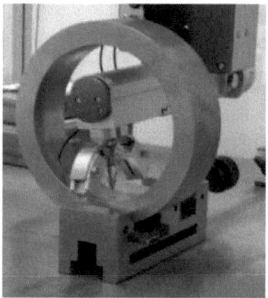

Figure 3.6: X-ray diffractometry for measuring residual stresses in a non-destructive way at the MCL (Materials Center Leoben, Forschung GmbH). (courtesy of G. Figala, LUT, Montanuniversitaet Leoben)

Chapter 4

Laser Treatment of Plate-Structures

After a short introduction, the current Chapter gives a description of the applied FEM models for laser treatment of plate-structures and the performed simulations. Firstly the development of the temperature and stress fields during the heat input are numerically investigated. As second step different geometric laser tracks and laser point patterns are considered and evaluated with respect to their effect on buckling load increase as well as eigenfrequency enhancement in comparison to untreated plate-structures. Further investigations are carried out on the stiffening effect of laser treatment (buckling stiffness, eigenfrequency increase) of individual shear webs in framed structures, representing a situation rather prevailing in the application of plate-structures, e.g., in railway car body design. The Chapter is closed by showing some experimental results in comparison to the simulation results, and by a summary. Main parts of this Chapter are already published in [12] and [13].

4.1 Introduction

Thermal treatment is a widely used method in mechanical engineering to enhance different characteristics of components or products. Representing a classical method of thermal treatment with a vast number of applications, tempering of steel should be mentioned here, see eg. [52].

Distortions during heat treatment play an important role and can even result in a destruction of the component subjected to this kind of treatment. While tempering of steel requires a change in the material's micro structure, i.e., a phase change, i.e., from Ferrite- (α-Fe) to Austenite- (γ-Fe) and finally to Martensite structure, specific thermal treatments may also act on other physical conditions of the components subjected to the respective treatment. A sometimes desired, as well as in other cases undesired, effect of many thermal treatments is the introduction of residual stresses into a structure as already discussed in more detail in Chapter 3.3. For thermal processes (e.g., thermal cutting, welding, etc.), which also deal with adequately high temperatures, residual stresses are, therefore, present in the affected structures.

4.2 Modelling of Laser Heat Treatment of Plate-Structures by the Finite Element Method

As shown in [44] (this literature deals with laser welding where a state change from solid to liquid occurs which is even more difficult to model compared to heat treatment while the material stays in a solid state) the development of residual stresses in components subjected to laser treatment can be devided into different phases. As already mentioned before, for the modelling applied in this Chapter, the material remains in the solid state and, therefore, main effects to be reproduced by the model can be described in a simplified manner, depicted as following:

- Heating up resulting in an occurrence of thermally induced stresses
- Exceeding of the yield limit resulting in plastic deformations
- Cooling-down and development of residual stresses

Examples from literature dealing with such considerations, can be found, e.g., in [50].

4.2.1 Specifying the Model-Type

For simulating phenomena as described above by a finite element model both heat transfer and stress field development have to be solved. Generally speaking, the actual model has to be a nonlinear model due to the occurence of plasticity and other material nonlinearities. Geometric nonlinearities may occur due to severe thermal distortions. Applying ABAQUS FEM software from Dassault Systèmes Simulia, see [1], one has the choice to separate the calculation of the transient temperature field from the stress analysis. In this one-sided coupled analysis the outcome of the thermal analysis, i.e., the time dependent temperature field is used as input for the second model which calculates the stress field. This modelling technique is only possible if no feedback from the stress (and deformation) analysis appears onto the heat input. More exact is the fully coupled thermomechanical analysis solving both thermal- and stress problem simultaneously. A classical example requiring a fully coupled simulation approach is a model of a disc-brake where breaking conditions (temperature dependent friction and behaviour of the brake pads) as well as the resulting heat dissipation are directly influenced within a loop (see, e.g., ABAQUS Example Manual, [1]). In the case considered in this thesis a fully coupled modelling approach is chosen for convenience reason, although is not directly of physical necessity.

4.2.2 Material Data

For all numerical investigations in the present Chapter a mild steel type DC01 (according to standardisation EN 10130) is chosen as material (for reference purpose see, e.g., engineering specifications for delivery of a steel supplier [6]). Temperature dependent material data, which is necessary for the coupled thermo-mechanical analysis, is taken from [51]. Following physical quantities, all being temperature dependent, are used in the simulation, see Figs. 4.1 and 4.2:

- Mass Density ρ
- Specific Heat Capacity c_p
- Thermal Conductivity λ
- Coefficient of Thermal Expansion α
- Young's Modulus E
- Poisson's Ratio ν
- Yield Strength σ_F

Plastic hardening is presented by hot tensile tests performed at the LUT in Leoben, depicted in Fig. 4.2.

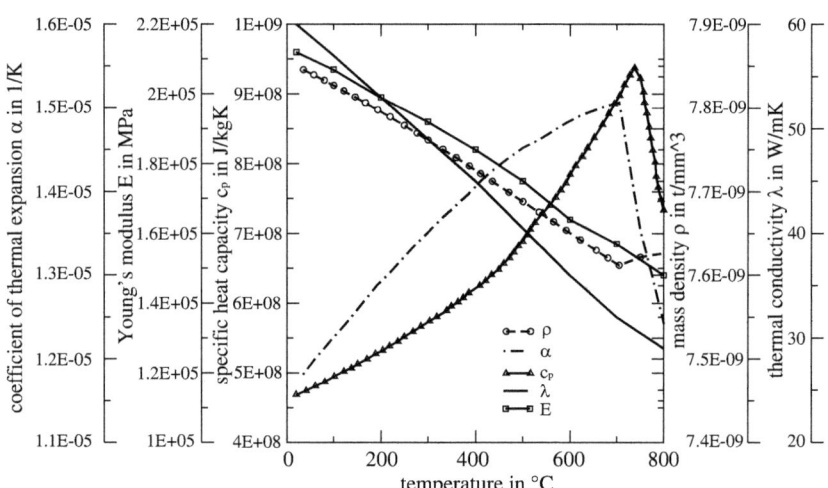

Figure 4.1: Material properties of mild steel type DC01 over a temperature range up to 800°C, taken from [51].

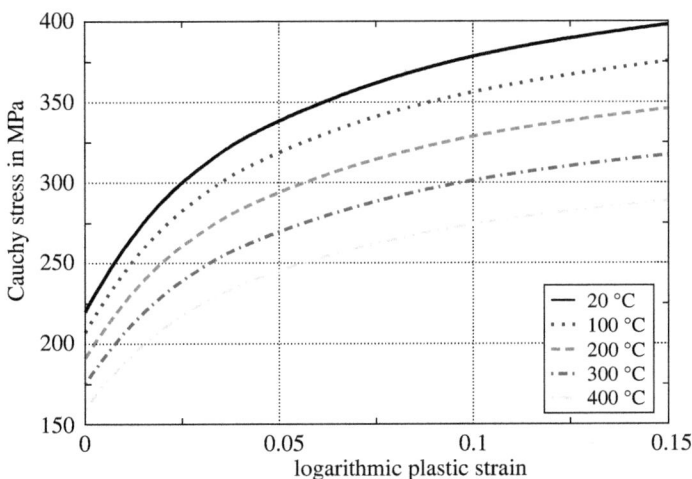

Figure 4.2: Material flow curves of mild steel type DC01 at specific temperatures due to hot tensile tests performed at the LUT in Leoben. (courtesy of G. Figala, LUT, Montanuniversitaet Leoben)

4.2.3 Modelling of the Laser Heat Input

A three dimensional modelling approach is chosen for the FEM model. Due to the kind of structures which are to investigate (large dimensions compared to the thickness), structural elements (shell elements - in FEM nomenclature) are chosen for discretisation in the FEM model. For reference purpose see for example [16]. Linear S3T- (ABAQUS, [1], manual nomenclature: '3-node triangular general-purpose shell, finite membrane strains, bilinear temperature in the shell surface') and S4T-elements (ABAQUS, [1], manual nomenclature: '4-node doubly curved general-purpose shell, finite membrane strains, bilinear temperature in the shell surface') are applied. These elements have in addition to displacement and rotation degrees of freedom also a temperature degree of freedom. Due to the very local acting of the laser beam a finer element discretisation has to be introduced into the model along the path the laser moves or at laser points were laser irradiation takes place. Mesh details are shown in forthcoming figures.

Heat introduction in the simulation model is being done by applying a 'nonuniform distributed flux', as it is called according to ABAQUS nomenclature, onto the top element surface (with respect to the shell normal vector). In [9], a Gauss-distributed heat source is proposed for laser heat flow intensity input. Equation 4.1 represents an axisymmetric, uncut and Gauss-distributed heat flow intensity with x', y', z' representing the position in a Cartesian coordinate system at time t with the origin of the coordinate system in the midpoint of the laserfocus and a focus radius r_F. The absorption coefficient of the structure's surface is given by A_L, and P_L is the laser power which would be introduced into the structure if an uncut Gauss-distribution was utilised. Further components of the equation are the Dirac distribution $\delta(z')$ and the Heavyside function $\tau(t')$. Compared to the original representation, given in [9], in Eq. 4.1 the multiplicative factor '4' was changed to factor '2' because in the current model only the irradiation submitted into the halfspace (structure) is of interest.

$$Q_w(x', y', z', t') = \frac{2 A_L P_L}{\pi r_F^2} e^{-2\left(\frac{x'^2}{r_F^2} + \frac{y'^2}{r_F^2}\right)} \delta(z') \tau(t') \tag{4.1}$$

Practical investigations conducted by the LUT at the MUL in Leoben, see details in report [25], show a sharp zoning of the occuring annealing colours on plate surfaces subjected to laser irradiation. Because of this observation the Gauss-distribution of the heat flow intensity, used in the majority of the analyses, is cutted along the focus radius, i.e., assuming that for $r > r_F$ no heat is transferred to the surface. In the cutted case an integration over the remaining area shows that only 86 %, compared to the energy available by an uncut Gauss-distribution, is available.

Thus, this calculated heat per area and time unit (under consideration of the absorption coefficient A_L) is introduced in the FEM model as a position and time dependent heat source, acting on the structure's surface which is facing the laser beam. For this purpose an auxiliary programme in the FEM model, in ABAQUS nomenclature a subroutine, is being applied. Effects of reflexion and radiation at the surface are already taken into account by the absorption coefficient. Convective heat transfer is used on both of the structure's surfaces to represent a heat sink.

4.2.4 Development of Residual Stresses, FEM Investigations of Laser Heat Treatment with a single Laser Point

In this section the development of stresses during single point laser heat treatment and the finally remaining residual stresses after the treatment are investigated numerically. These investigations are performed applying the already described FEM modelling strategies. A comparison with results obtained from accompanying experimental investigations, see [23] and [41], is given at the end of this section.

In the current simulations heat input is introduced by a single laser point at the midpoint of a simply-supported square plate with $a_0 = 90$ mm, see Fig 4.3. The boundary conditions are present while laser irradiation is activated (heat input) as well as during the cooling down period, and no additional external load is present at any time. The influence of two main parameters is investigated, the plate thickness and the duration of laser irradiation. The plate's thickness is chosen between 0.75 mm and 2 mm. For the laser irradiation duration values between 0.5 s and 1 s are chosen. For each subsequent analysis the parameter values are given in detail in the respective section.

A laser power $P = 515$ W is chosen while for the absorption coefficient on the plate's top-surface position $A_L = 0.3$ is selected. For objective investigations a trimming of the Gauss-distributed heat flow intensity at the focus radius r_F, as described in Chapter 4.2.3, is not employed. For all investigations a suitable duration of the

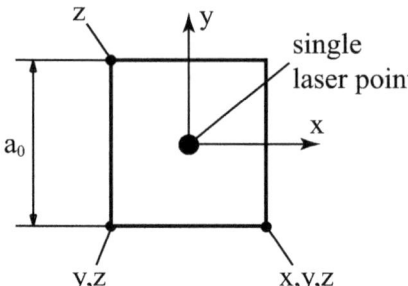

Figure 4.3: Boundary conditions of the square plates (treated with a single laser dot) with $a_0 = 90\,\text{mm}$ used in the FEM analyses to investigate the residual stress development during heat treatment.

cooling down time of 500 s is chosen, leading to a stationary situation at the end of the simulation.

For all forthcoming investigations especially of interest are the σ_{yy} and the σ_{xx} stress components in the plate's mid plane. These stress components give evidence of possible enhancement if an external load is applied in one of those directions.

Chapter 4 Laser Treatment of Plate-Structures

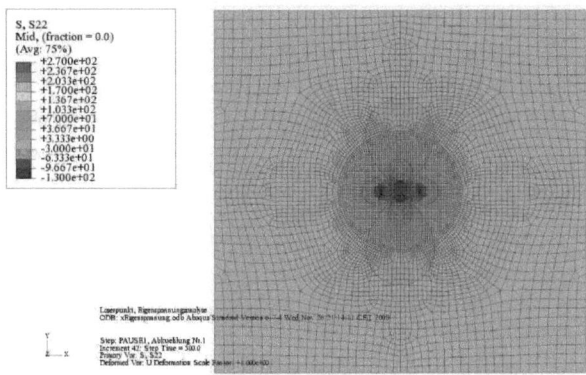

Figure 4.4: Fringe-plot showing the distribution of σ_{yy}-stresses in MPa of plate type W2B1.1, treated 0.5 s with a single laser point. Depicted stress state is present in the plate's mid-plane after complete cooling down.

Investigated Plate Structure Type W2B1.1

First plate-structure investigated is denoted as type W2B1.1 and has a thickness of 0.75 mm. A laser irradiation duration of 0.5 s is used for the FEM analysis. The distribution of σ_{yy}-stresses in the plate's mid-plane (after complete cooling down) is depicted as a fringe plot in Fig. 4.4.

A comparison of σ_{yy}-distributions (plate's top-surface, plate's mid-plane and plate's bottom surface) along a horizontally advancing path is shown in Figs. 4.9, 4.10 and 4.11. Stress distributions obtained on the plate's top-surface, in the plate's mid-plane and on the plate's bottom-surface, respectively, show a slightly different behaviour which can be explained due to the unsymmetric, one-sided heat input.

Development and respective distributions of σ_{xx} as well as σ_{yy}-stresses during the heat input (horizontal path running through the plate's midpoint, stress values are evaluated at the top surface which is subjected to heat treatment) are depicted in Fig. 4.5 as well as in Fig. 4.7. The development of σ_{xx} and σ_{yy}-stresses during the cooling down and the resulting residual stress state, respectively, are shown in

Chapter 4 Laser Treatment of Plate-Structures

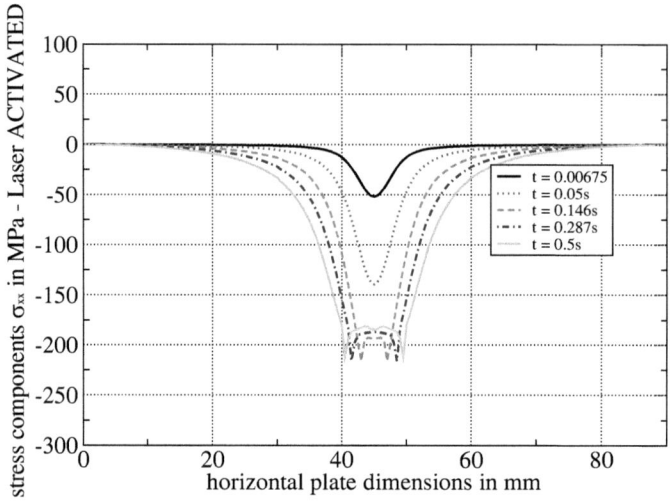

Figure 4.5: Horizontal distribution of σ_{xx}-stresses in MPa, plate's top-surface position, while laser irradiation is activated (heat input), plate type W2B1.1.

Figs. 4.6 and 4.8. A remarkable fast rearrangement of the stresses is especially observable for the begin of heating-up and also the set in of the cooling down phase.

Chapter 4 Laser Treatment of Plate-Structures

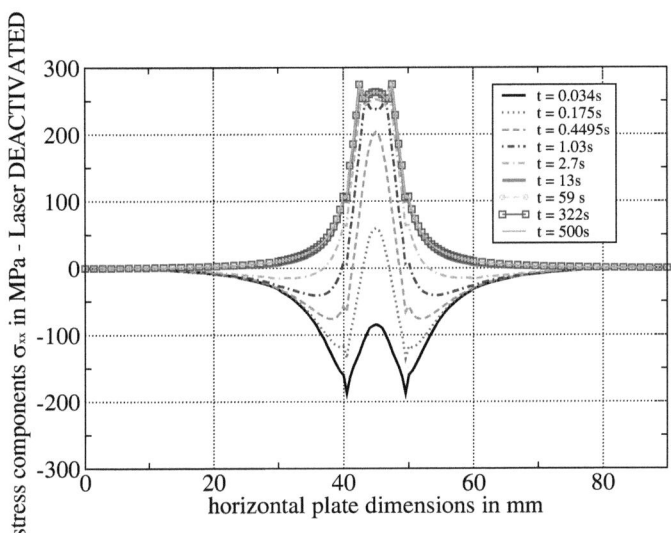

Figure 4.6: Horizontal distribution of σ_{xx}-stresses in MPa, plate's top-surface position, during cooling down, plate type W2B1.1.

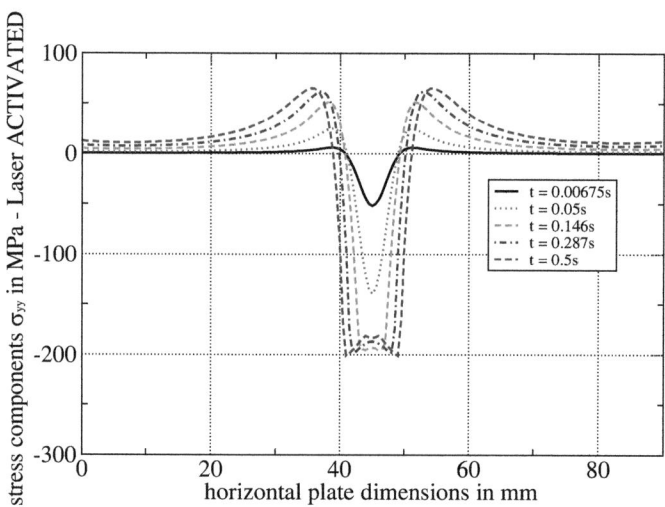

Figure 4.7: Horizontal distribution of σ_{yy}-stresses in MPa, plate's top-surface position, while laser irradiation is activated (heat input), plate type W2B1.1.

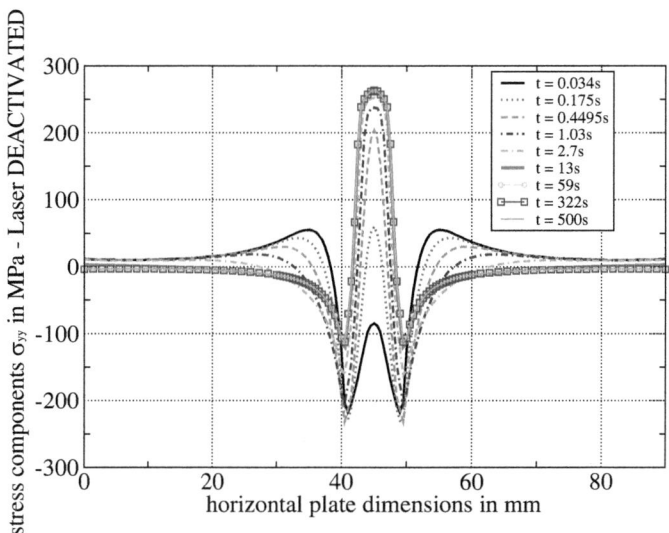

Figure 4.8: Horizontal distribution of σ_{yy}-stresses in MPa, plate's top-surface position, during cooling down, plate type W2B1.1.

Chapter 4 Laser Treatment of Plate-Structures

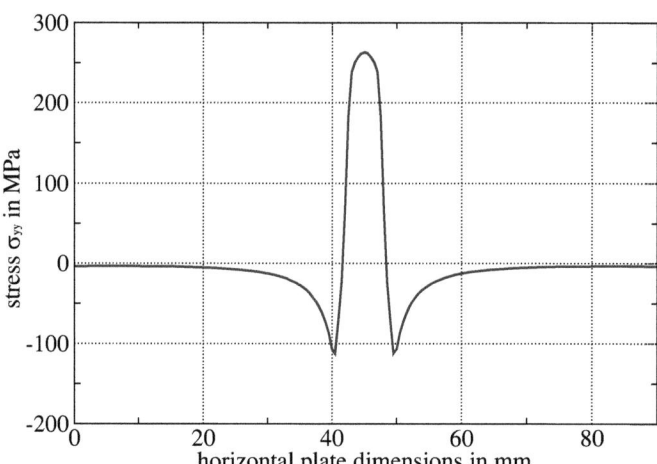

Figure 4.9: σ_{yy}-stress distribution in MPa, along a horizontal path, on the plate's top-surface. The horizontal path advances through the plate midpoint, plate type W2B1.1.

Chapter 4 Laser Treatment of Plate-Structures

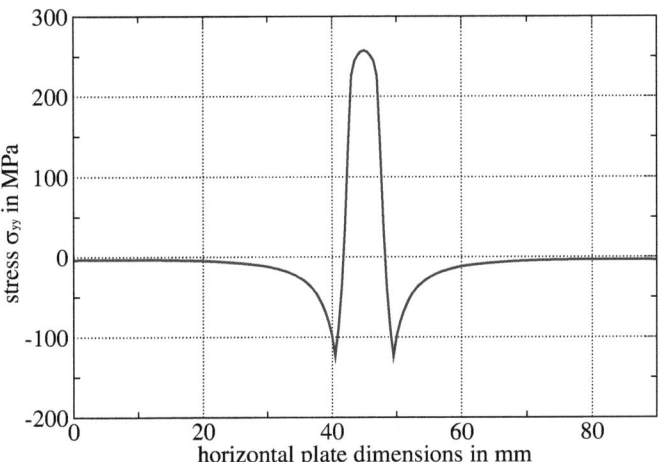

Figure 4.10: σ_{yy}-stress distribution in MPa, along a horizontal path, in the plate's mid-plane. The horizontal path advances through the plate midpoint, plate type W2B1.1.

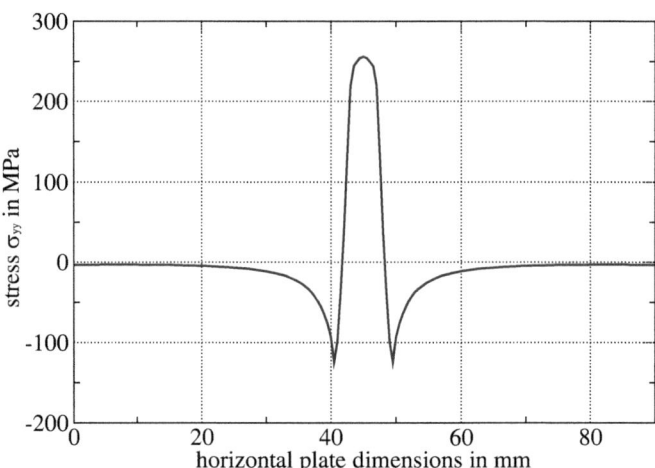

Figure 4.11: σ_{yy}-stress distribution in MPa, along a horizontal path, on the plate's bottom surface. The horizontal path advances through the plate midpoint, plate type W2B1.1.

Chapter 4 Laser Treatment of Plate-Structures

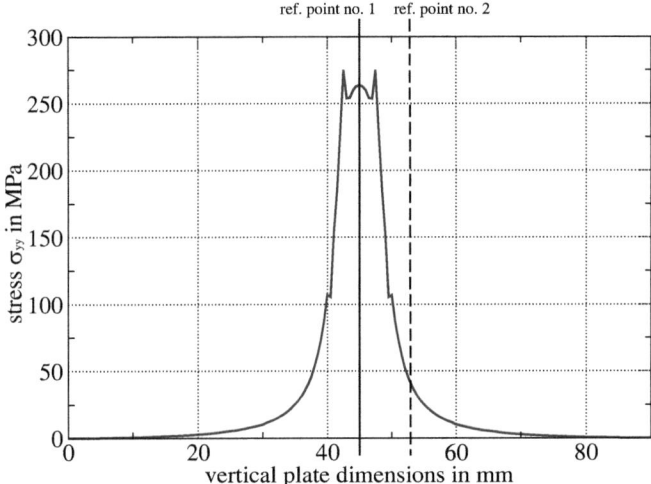

Figure 4.12: σ_{yy}-stress distribution in MPa, along a vertical path, on the plate's top-surface. The vertical path advances through the plate's midpoint, plate type W2B1.1.

Figures 4.12, 4.13 and 4.14 show a compilation of σ_{yy}-stress distributions each along a vertical path which advances through the plate's mid-point on the plate's top-surface, in the plate's mid-plane and on the plate's bottom surface, respectively. One can observe a nearly constant stress distribution over the plate's thickness.

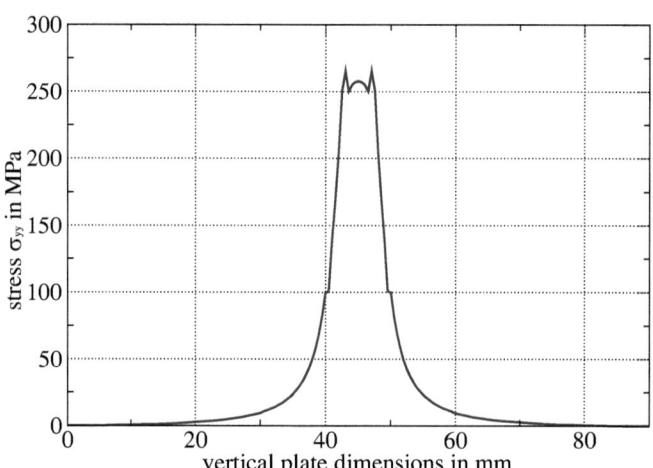

Figure 4.13: σ_{yy}-stress distribution in MPa, along a vertical path, in the plate's mid-plane. The vertical path advances through the plate's midpoint, plate type W2B1.1.

Chapter 4 Laser Treatment of Plate-Structures

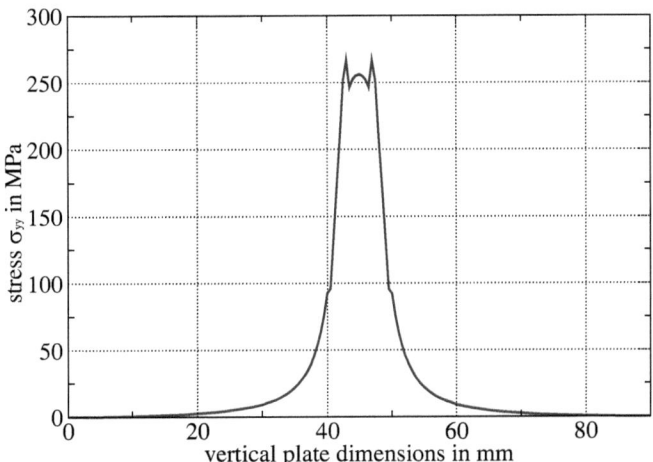

Figure 4.14: σ_{yy}-stress distribution in MPa, along a vertical path, on the plate's bottom surface. The vertical path advances through the plate's midpoint, plate type W2B1.1.

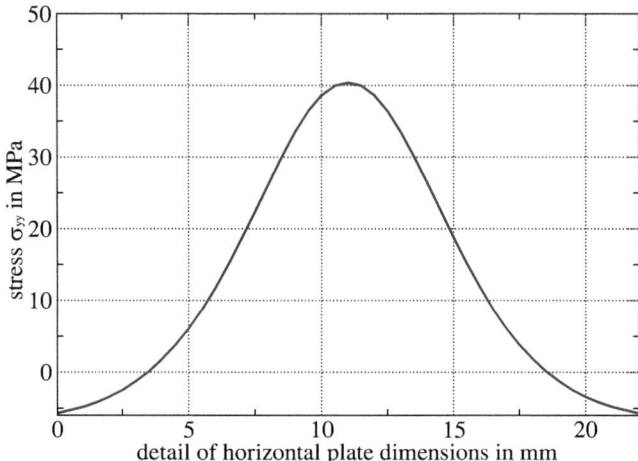

Figure 4.15: Details of σ_{yy}-stress distribution, along a horizontal path on the plate's top-surface. The horizontal path advances through reference point no. 2, plate type W2B1.1.

For comparison purpose the position of two reference points, no. 1 and no. 2, is denoted in Fig. 4.12. Reference point no. 2 is situated about 8 mm below reference point no. 1 which corresponds to the laser midpoint (in current case the laser midpoint is the plate's midpoint).

In Figs. 4.15, 4.16 and 4.17 the horizontal σ_{yy}-stress distribution, running through reference point no. 2 is shown. Evaluation is done, in this case only, for a path of ca. 22 mm symmetric to the plate's y-axis because locally situated but advancing mesh refinement heading radially towards the laser midpoint results in partially inclined elements, leading to a non-even stress path.

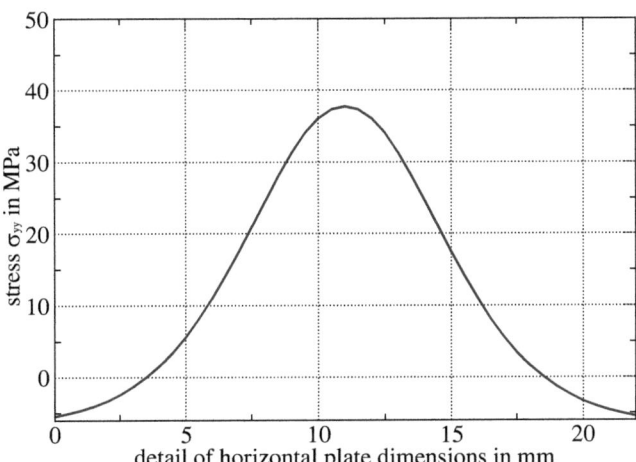

Figure 4.16: Details of σ_{yy}-stress distribution, in the plate's mid-plane. The horizontal path advances through reference point no. 2, plate type W2B1.1.

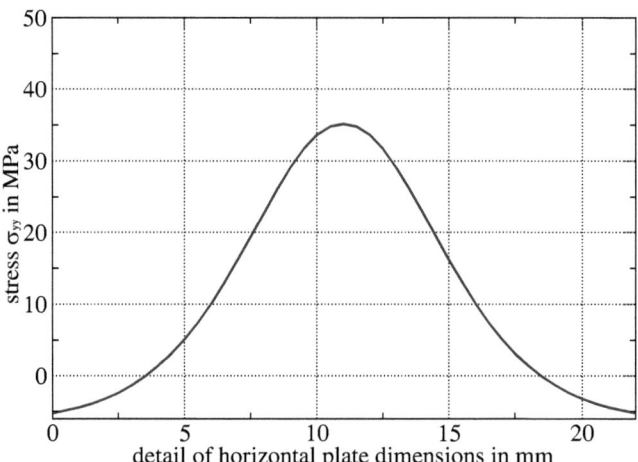

Figure 4.17: Details of σ_{yy}-stress distribution, along a horizontal path on the plate's bottom surface. The horizontal path advances through reference point no. 2, plate type W2B1.1.

Chapter 4 Laser Treatment of Plate-Structures

Results of the σ_{yy}-stress distribution in reference point no. 1, along the plate's thickness coordinate, are shown in Fig. 4.18. As a direct comparison, in Fig. 4.19 the σ_{yy}-stress distribution along the thickness coordinate in reference point no. 2 is shown.

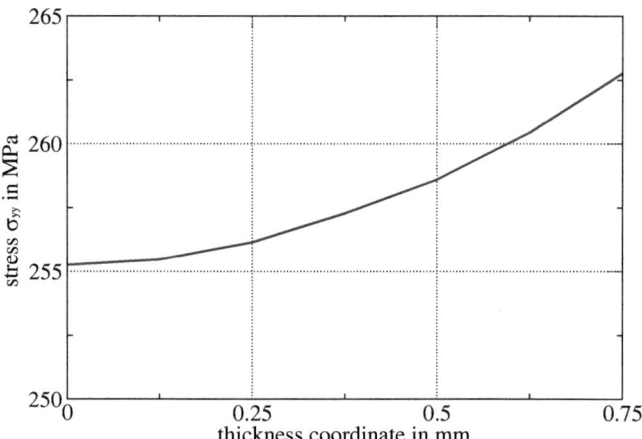

Figure 4.18: Distribution of σ_{yy}-stresses in MPa along the plate's thickness coordinate in reference point no. 1 (plate's midpoint), plate type W2B1.1.

Chapter 4 Laser Treatment of Plate-Structures

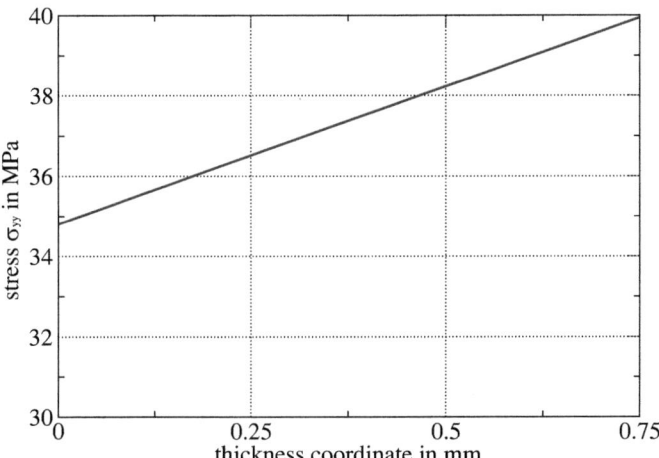

Figure 4.19: Distribution of σ_{yy}-stresses in MPa along the plate's thickness coordinate in reference point no. 2, plate type W2B1.1.

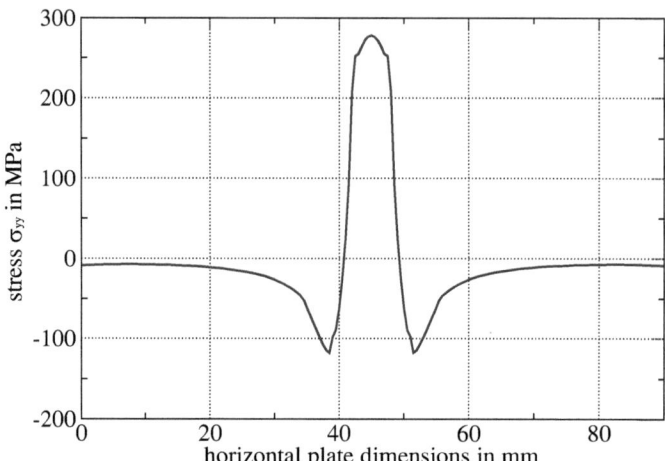

Figure 4.20: σ_{yy}-stress distribution in MPa, horizontal path, on the plate's top-surface, plate type W2B1.2.

Investigated Plate-Structure Type W2B1.2

For this structure the duration of laser irradiation is increased to 1 s, all other parameters are the same as for structure type W2B1.1. Stress distributions for σ_{yy} (horizontally) as well as for σ_{xx} (vertically) are depicted in Figs. 4.20, 4.21, 4.22 and Figs. 4.23, 4.24, 4.25 respectively.

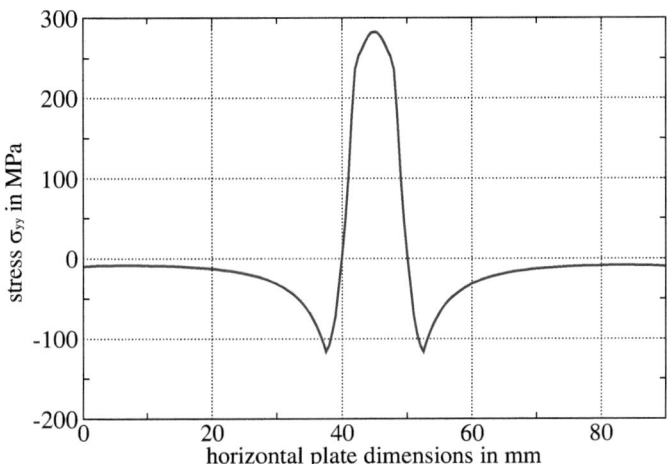

Figure 4.21: σ_{yy}-stress distribution in MPa, horizontal path, in the plate's mid-plane, plate type W2B1.2.

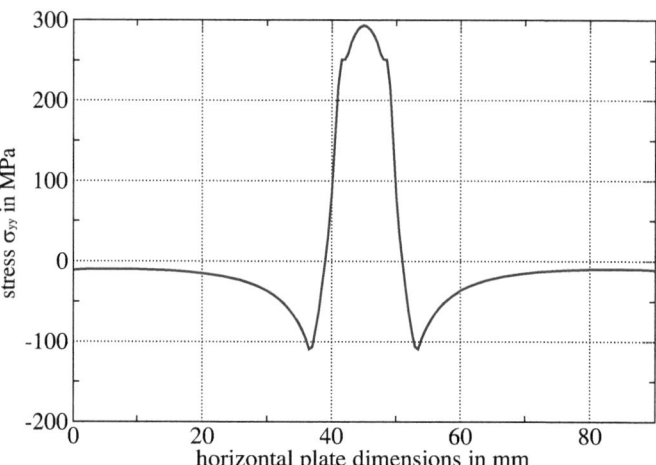

Figure 4.22: σ_{yy}-stress distribution in MPa, horizontal path, on the plate's bottom surface, plate type W2B1.2.

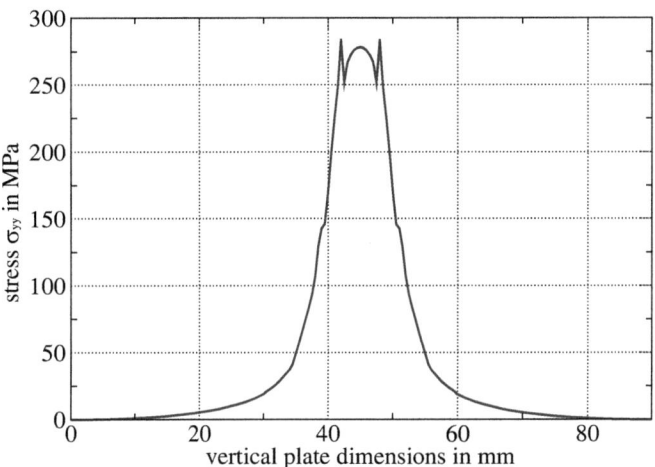

Figure 4.23: σ_{yy}-stress distribution in MPa, vertical path, on the plate's top-surface, plate type W2B1.2.

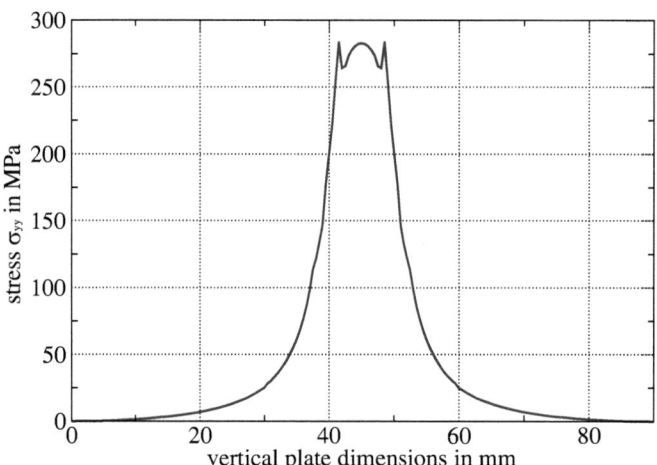

Figure 4.24: σ_{yy}-stress distribution in MPa, vertical path, in the plate's mid-plane, plate type W2B1.2.

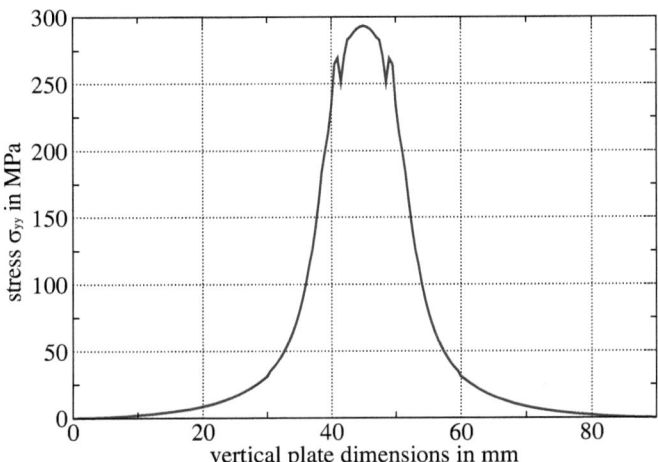

Figure 4.25: σ_{yy}-stress distribution in MPa, vertical path, on the plate's bottom surface, plate type W2B1.2.

Chapter 4 Laser Treatment of Plate-Structures

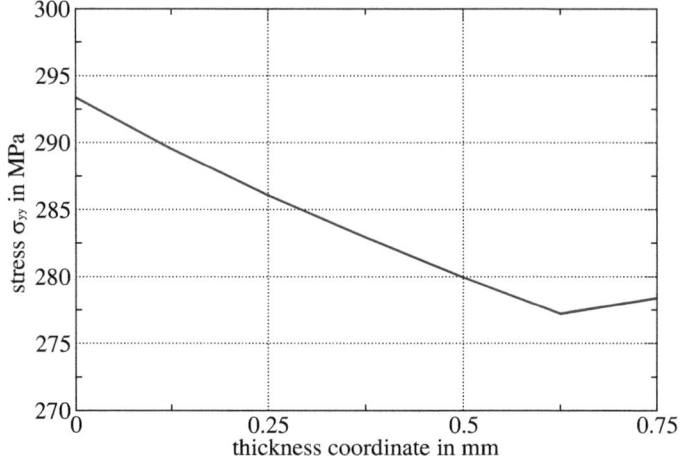

Figure 4.26: σ_{yy}-stress distribution in MPa, along the plate's thickness in the plate's midpoint, plate type W2B1.2.

An investigation of the σ_{yy}-stress distribution through the plate's thickness (position of the plate's midpoint) shows the characteristic development, as depicted in Fig. 4.26. A change in the inclination of the stress value curve, in its first third, is noticeable. One reason for this characteristic, in comparison to results obtained for plate W2B1.1, might be the emerging of larger deformations (deformation magnitude of ca. 0.5 mm for plate type W2B1.2 vs. ca. 0.0085 mm for plate type W2B1.1, each in the situation of maximum prevailing temperatures) while heat treatment. However the maximal remaining deformation magnitude (0.06 mm) for the completely cooled-down plate W2B1.2 is still small (maximal remaining deformation magnitude plate type W2B1.1 ca. 0.009 mm).

Chapter 4 Laser Treatment of Plate-Structures

Investigated Plate Structure Type W2B4.1

Structure type W2B4.1 represents with 2 mm the thickest plate metal investigated. In current case the applied duration of laser irradiation is 0.5 s. Horizontal stress distributions for σ_{yy}-stresses are depicted in Figs. 4.27, 4.28 and 4.29. Not only with respect to the quality, but also to the quantity these σ_{yy} residual stress distributions in each plane (plate's top-surface, plate's mid-plane, and plate's bottom surface) show considerable differences which, in this case, might come from more uneven temperature fields in the quite thick plate. Stress distributions through the plate's thickness, displayed in Fig. 4.30, show, in comparison to above described thinner plates, less tensile stresses. These tensile stresses decrease further during advancing in the thickness coordinate to the plate's centre until reaching nearly zero. On the plate's bottom surface, which is the surface averted to the laser irradiation, again tensile stresses are present (even though they are rather small).

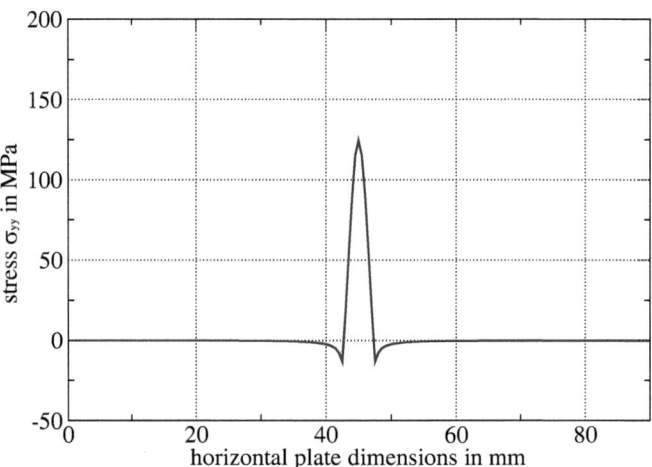

Figure 4.27: σ_{yy}-stress distribution in MPa, along a horizontal path, on the plate's top-surface, plate type W2B4.1.

Chapter 4 Laser Treatment of Plate-Structures

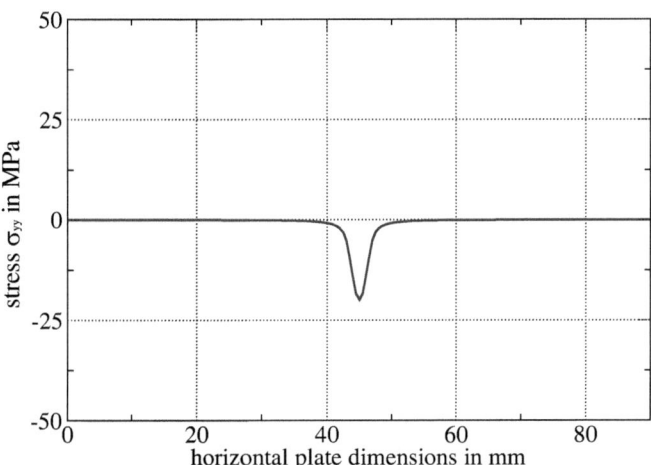

Figure 4.28: σ_{yy}-stress distribution in MPa, along a horizontal path, in the plate's mid-plane, plate type W2B4.1.

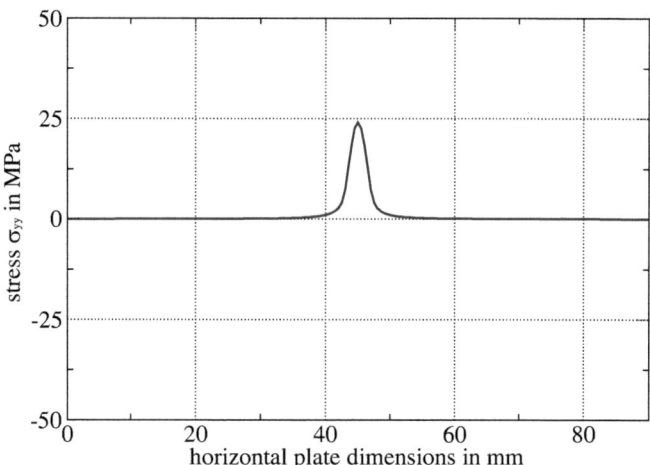

Figure 4.29: σ_{yy}-stress distribution in MPa, along a horizontal path, on the plate's bottom surface, plate type W2B4.1.

Chapter 4 Laser Treatment of Plate-Structures

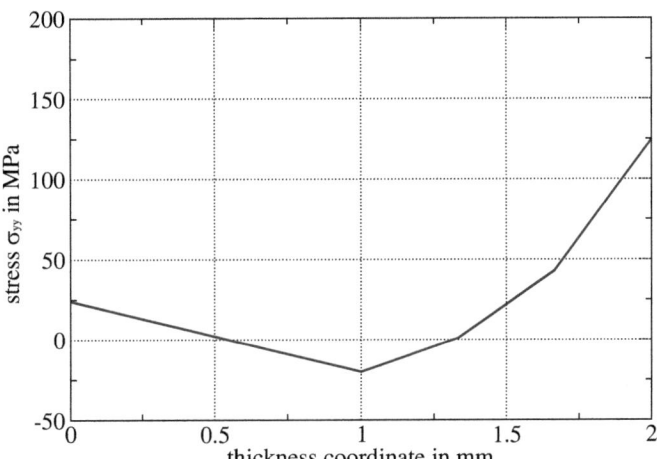

Figure 4.30: σ_{yy}-stress distribution in MPa, along the plate's thickness in the plate's midpoint, plate type W2B4.1.

Chapter 4 Laser Treatment of Plate-Structures

Comparison of FEM-Simulation Results to Experimental Results

FEM simulations show for all investigated plate-structures (type W2B1.1, W2B1.2, and W2B4.1), despite different parameters, similar qualitative results concerning stress development within the laser influenced zone.

Concerning stability (buckling stability) and vibration behaviour, respectively, membrane forces per unit length, contained within the plate, are decisive. Thus, measuring residual stresses along the plate's thickness is of special interest. This allows to investigate a possible gradient of residual stresses as well as a possibly existing bending stress state. Results obtained from measurements, see [41], using X-ray diffractometry show partially extensive deviations, compared to results obtained by simulations. These deviations are both, in magnitude and in the kind of the measured stress (tensile or pressure stress). It has to be stated, that the X-ray diffractometry method has only a very small depth range. Therefore explanations for the deviations could be surface effects (eventually present due to production steps like roll forming or other mechanical surface treatments). Applying the destructive borehole method, which partially allows a detailed measurement over a broader range of material thickness, see [23] or in general [45], is the better choice in the current case.

Another noteworthy observation, obtained by simulation results, is the reproduction of laser induced local deformations showing a cone-shaped depression in the plate-structure, see Fig. 4.31. Local deformations from that kind result in changes of the stress distribution along the plate's thickness compared to an, in most cases, nearly linear stress distribution as it is present outside of the heat affected zone (HAZ), compare Figs. 4.18 and 4.19.

4.3 Analysis of Laser Treated Plates

As it is shown in [46] an appropriate distribution of residual membrane stresses in a plate is able to increase it's buckling resistance as well as enhances it's frequency behaviour (for example in terms of an increase of the fundamental frequency).

Until now, all investigated plate-structures, see, e.g., Fig. 4.3, were unloaded. The character of the exhibition of residual stresses was the main focus point of the

Chapter 4 Laser Treatment of Plate-Structures

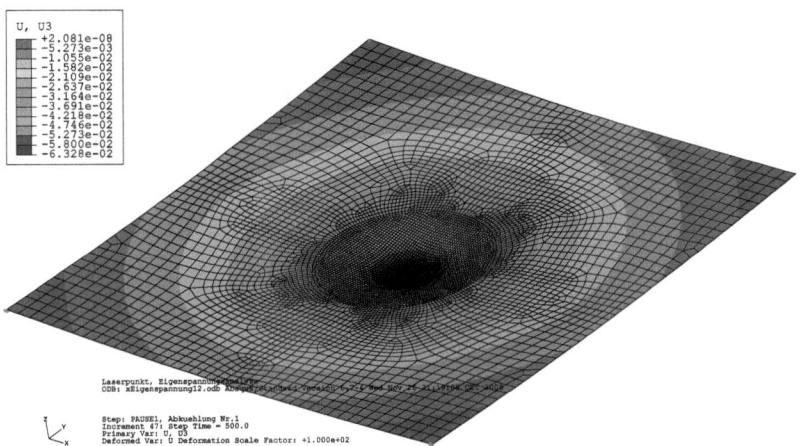

Figure 4.31: Displacement normal to plate-plane U3 (direction z-axis) in mm, existing in the plate after heat input and complete cool-down. (Deformation scale factor: 100), plate type W2B1.2.

Chapter 4 Laser Treatment of Plate-Structures

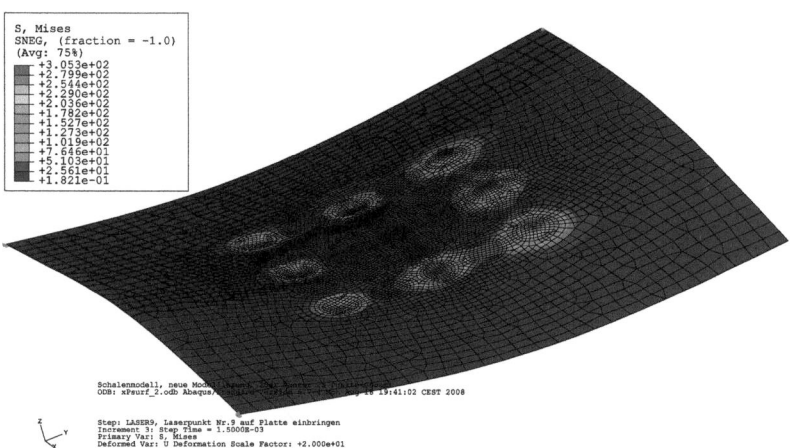

Figure 4.32: Simulation showing a 'foldover'-like behaviour with deformation direction changing during laser irradiation. For enhancement of visibility reason fringe plots of von Mises equivalent stresses in MPa (which exhibit high values within the laser influencing area) are shown (situation on the plate's bottom surface). (Deformation scale factor: 20).

Chapter 4 Laser Treatment of Plate-Structures

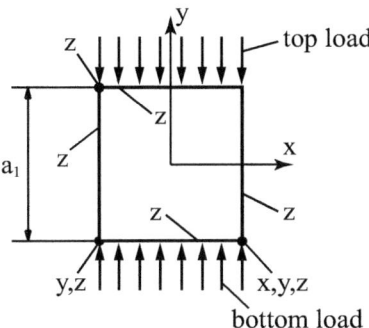

Figure 4.33: Sketch of the general loading situation of the laser treated square plate-structures (dimensions: $a_1 = 280$ mm, thickness 0.75 mm). Denoted directions x, y or z, with the plate lying in the x,y-plane, show a blockage in respective translatoric direction.

investigations. For upcoming analysis, larger plate-structures are investigated (dimensions: $a_1 = 280$ mm, thickness 0.75 mm) which are subjected to boundary and loading conditions as depicted in Fig. 4.33. According to the shown loading conditions for the larger plate-structures, the exhibition of the σ_{yy}-stresses, with respect to their membrane stress fraction, is of special interest for positively enhancing the plate-structure's characteristics, as already statet before.

An introduction of such an appropriate distribution of stresses is however a challenging task. Laser irradiation, featuring a very local but highly intensive heat source, is the method of choice. The already mentioned square plate-structures, which exhibit an edge length of $a_1 = 280$ mm and a thickness of 0.75 mm (see sketch in Fig. 4.35), are also chosen because of plate-measures being still suitable for the testing machine situated at the LUT in Leoben.

An untreated plate with given dimensions is furthermore denoted as standard plate. A general distinction between continuously laser treated plates and pointwise laser treated plates is done in the following. Generic models being used feature a circle-, an I-, and an X- laser track geometry for continuous laser irradiation. Additionaly,

Chapter 4 Laser Treatment of Plate-Structures

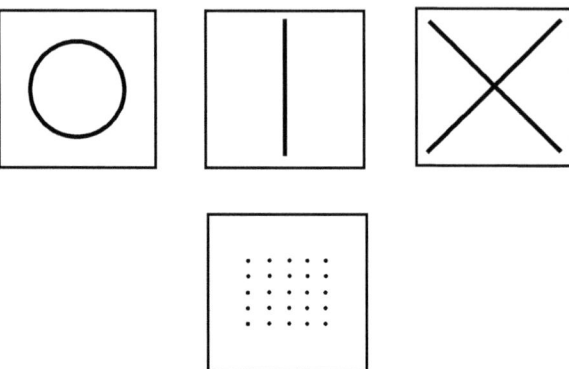

Figure 4.34: Sketch of introduced laser track geometries and point pattern. Continuous circle-, I- and X- geometry as well as point pattern for the pointwise laser treatment.

generic models for different laser point patterns are used. In Fig. 4.34 a sketch of each basic laser track- and point pattern geometry is shown.

4.3.1 Continuously Laser Treated Plates

For continuous laser treatment a trimmed Gauss-distributed heat flow intensity, as explained in Chapter 4.2, is applied. The laser track width is 9 mm, being twice the focus radius. Laserpower P_L is chosen to be 400 W, the absorption coefficient $A_L = 0.3$. For the focus-radius r_F, a value of 4.5 mm is used. According to these choices the relevant laser power P^* for heating results in the amount given in Eq. 4.2.

$$P^* = 0.86 \cdot P_L \cdot A_L = 0.86 \cdot 400W \cdot 0.3 = 103.2W \qquad (4.2)$$

For the models featuring continuous laser treatment a movement speed of the laser beam is chosen to be $1\,\mathrm{m/min}$ ($16.66\,\mathrm{mm/s}$).

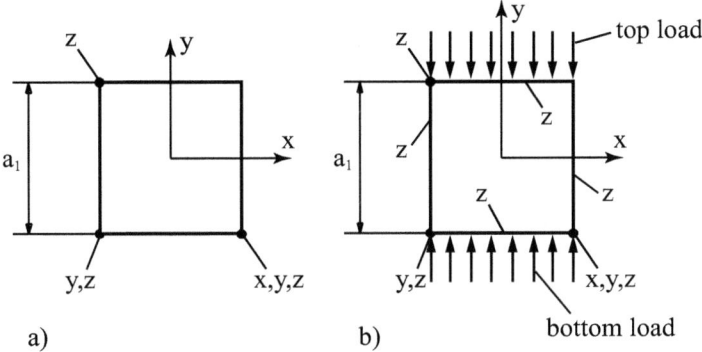

Figure 4.35: Boundary conditions of the square plates with $a_1 = 280\,\text{mm}$ and a thickness of $0.75\,\text{mm}$. a) Boundary conditions used during the FEM analyses for the laser heat input and cooling down. b) Boundary and loading conditions being active during buckling- and eigenfrequency analyses. Denoted directions x, y or z, with the plate lying in the x,y-plane, show a blockage in respective translatoric direction.

Boundary Conditions

Different boundary- and loading conditions are present during laser heat input and during the buckling- and eigenfrequency analyses. In Fig. 4.35 boundary- and loading conditions are depicted for the particular analysis. Denoted directions x, y or z, with the plate lying in the x,y-plane, show a blockage in respective translatoric direction.

The models partially exhibit large heat induced deformations during the laser heat input simulation (boundary conditions as depicted in Fig. 4.35 case a) are active). For the successive analyses boundary conditions are changed to the situation shown in Fig. 4.35 case b) (simply supported boundary conditions with all edges blocked in direction z). These new boundary conditions are either directly applied onto the deformed plate-structure or an edge 'pullback', back to original zero position

Chapter 4 Laser Treatment of Plate-Structures

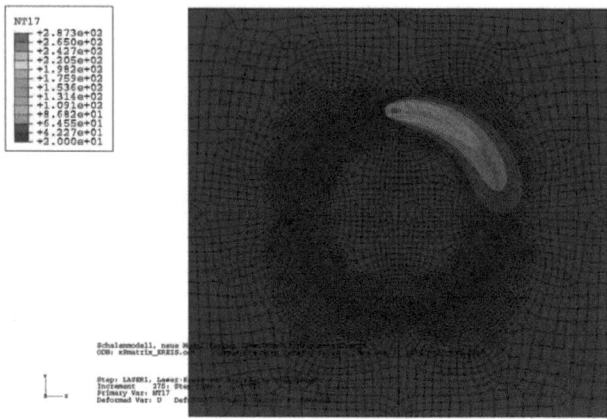

Figure 4.36: Fringe plot of temperature field in °C at the plate's top-surface, heat input due to moved laser source, circle geometry.

in coordinate-direction z, is applied. Depending on the residual deformations and the stress state that exists after laser treatment, a substantial residual stress rearrangement can appear during such an edge 'pullback', even plastic deformations can thereby possibly occur.

Laser-Circle Geometry introduced into Plate

In this subsection plates treated by laser running along a circular path are considered. The circle has a mean radius of 70 mm with the centre of the circle at the plate's midpoint. Speed of the laser beam is, as already stated before, modelled with $16.66\,\mathrm{mm/s}$ (equal to $1\,\mathrm{m/min}$).

Figure 4.36 shows a fring plot of the temperature field at the plate's top surface (facing the laser beam). In this figure the laser has already completed about one quarter of the total circle track length. A difference of ca. 13 °C in the temperatures at the plate's top-surface (maximal temperatures) compared to the temperatures appearing at the plate's bottom-surface results from the simulation.

Chapter 4 Laser Treatment of Plate-Structures

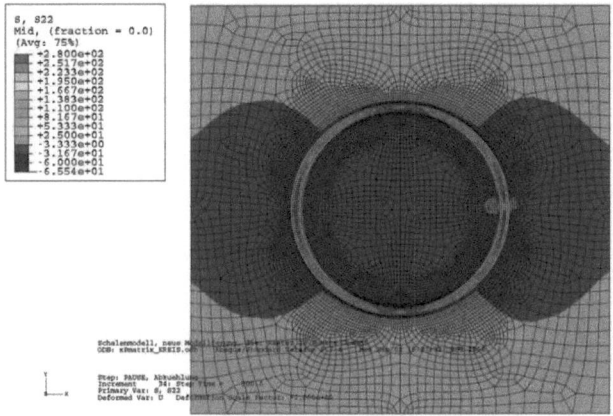

Figure 4.37: Fringe Plot of residual stresses σ_{yy} in MPa situated in the plate's mid-plane, heat input due to moved laser source, circle geometry.

After heating and complete cooling down to environmental temperature, which is simulated being 20 °C, resulting residual stresses σ_{yy} in the plate's mid-plane are shown as fringe plot in Fig. 4.37. An uneven distribution of residual stresses σ_{yy} at the laser track, existing at the 3 o'clock position of the circle track, can be explained by the activation and simultaneous movement of the laser source at the beginning of the heat introduction simulation. Due to this fact, the temperatures as well as stresses are different for start position (where they are considerably smaller) and some advanced position.

In Fig. 4.38 the residual stress distribution of σ_{yy}- stresses in the plate's mid-plane along a cutting path is depicted. This cutting path is running from 9 o'clock to 3 o'clock, horizontally through the treated plate's midpoint. Characteristic, fast decaying tensile stress peaks (when moving away from the immediate HAZ) are visible in the cut through the laser track. As already mentioned, at the 3 o'clock laser track position an uneven stress distribution is present, though also here a predominat tensile stress is visible but it has a decreased quantity compared to the one visible at for example the 9 o'clock position. During laser irradiation thermally

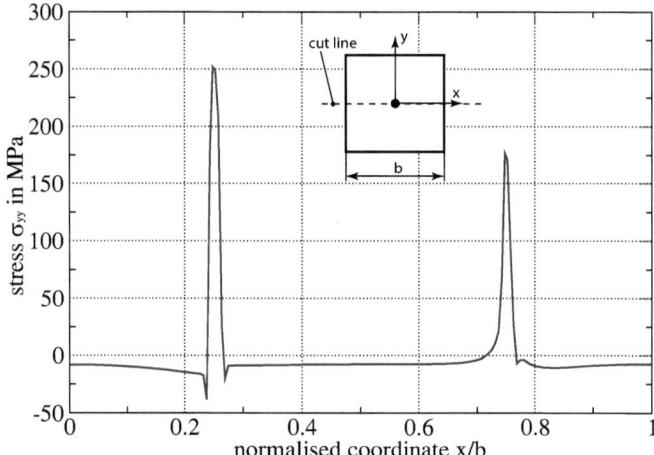

Figure 4.38: σ_{yy}-stress distribution in MPa, along $y \equiv 0$, after heat treatment with moving laser source, circle geometry. Stresses are situated in the plate's mid-plane.

induced deformations appear normal to the plate's plane. In Fig. 4.39 displacements in U3 (direction z) after cooling down are denoted as fringe plots. Highest residual displacements after complete cooling down are within the [mm] - range. These displacements along the horizontal path are shown in Fig. 4.40.

An illustration of the plate after completed laser treatment with the described circular laser pattern in isometric view under usage of a deformation scale of 10, see Fig. 4.41, shows clearly the finally existing deformations.

Laser-X Geometry introduced into Plate

For the X-geometry the laser track corresponds with the plate's diagonals but each laser track stops 10 mm before the plate's particular corner is reached. Figure 4.42

Chapter 4 Laser Treatment of Plate-Structures

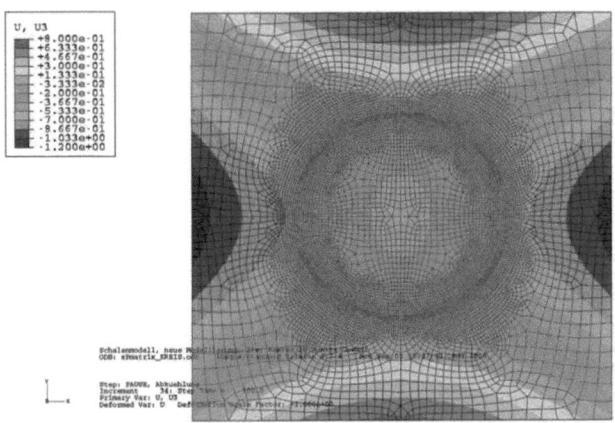

Figure 4.39: Fringe Plot of residual out-of-plane displacements U3 in mm. Heat treatment with moving laser source, circle geometry.

shows temperature fields at an instant during the introduction of the laser track no. 2. Also for this model a temperature difference between plate-top and plate-bottom surface is present, for maximal temperatures it is about 18 °C. All parameters concerning the laser modelling are identical to the model featuring laser track with circular geometry.

After laser treatment, along the X-shaped pattern, and complete cooling down to surrounding temperature (20 °C), a residual stress state exists, as shown in Fig. 4.43 for σ_{yy}.

Investigating stresses in the vicinity of the laser tracks (a cut is layed through the tracks), like for the laser treatment with circular geometry, σ_{yy}-tensile stress peaks are realisable. Stress distributions gained are not directly comparable to results from other modelled laser treated plates because of their horizontal cutting path compared to the inclined laser tracks which exist in the current case (inclination angle of 45°).

While the heat treatment in the FEM model is active severe deformations occur.

Chapter 4 Laser Treatment of Plate-Structures

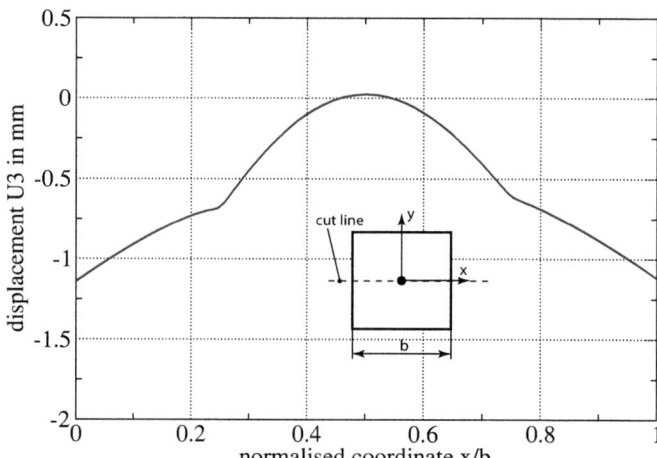

Figure 4.40: Residual out-of-plane displacement in mm along $y \equiv 0$. Heat treatment with moving laser source, circle geometry.

Chapter 4 Laser Treatment of Plate-Structures

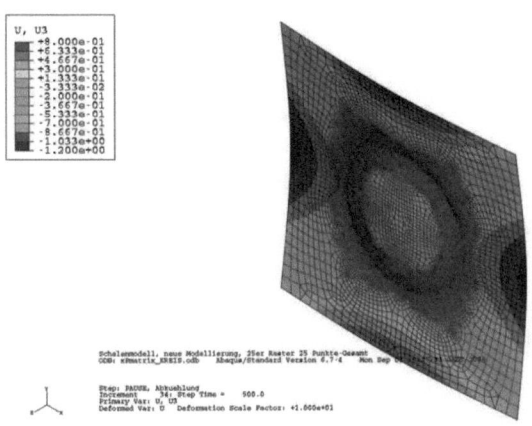

Figure 4.41: Fringe Plot of residual out-of-plane displacements U3 in mm. Isometric view, deformation scale 10. Heat treatment with moving laser source, circle geometry.

After complete cool-down residual out-of-plane deformations up to 8 mm remain. Figure 4.45 shows the remaining U3 deformations for the completely cooled-down plate. Especially at the plate's corner situated top right, which is free with respect to the boundary conditions, massive deformations are visible.

Figure 4.46, shows considerable out-of-plane displacements in negative z-direction where no support via boundary conditions is present. Figure 4.47 shows, by applying a deformation scale of 10, an isometric view of the U3 - deformations of the X-pattern laser treated plate.

Laser-I Geometry introduced into Plate

For the heat treatment according to an I-pattern geometry, the laser runs along a single vertical track which starts and ends 10 mm away from the respective plate edge to avoid edge influences. Again all laser relevant parameters are the same as used in the models for circle- and X-laser tracks before. Temperatures occuring while the laser is irradiating on the surface are depicted as fringe plot in Fig. 4.48.

Chapter 4 Laser Treatment of Plate-Structures

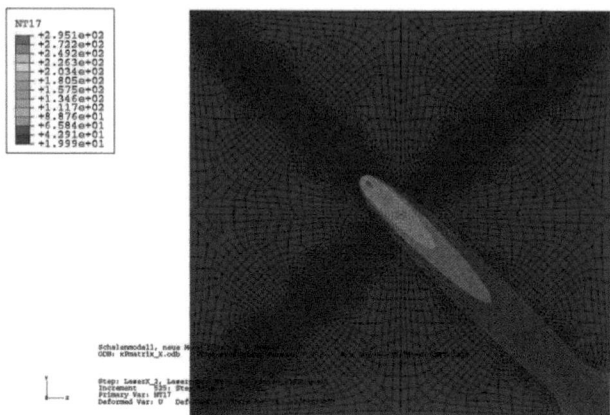

Figure 4.42: Fringe plot of temperature field in °C situated at the plate's top-surface, heat input due to moved laser source, X-pattern geometry.

Also the 16 °C temperature-difference between plate top and plate bottom surface, in the case at hand, is very similar to the already discussed models with continously acting laser irradiation.

Figure 4.49 shows the distribution of the residual stress component σ_{yy} at the plate's mid surface. And Fig. 4.50 shows again dominant tensile residual stresses σ_{yy} within the laser treated area. Residual deformations of the plate treated by an I-pattern laser geometry are depicted in Fig. 4.51 as fringe plot and in Fig. 4.52 as diagram for $y \equiv 0$. Highest occuring values for the residual deformations are about 1.5 mm.

The isometric view in Fig. 4.51 shows the residual displacements which are similar to the form of an open book.

4.3.2 Pointwise Laser Treated Plates Type 25P

Representing a discontinuous, pointwise laser treatment for the simulation type 25P a 5 x 5 points raster is laser treated. A distance of 25 mm is present between the

Chapter 4 Laser Treatment of Plate-Structures

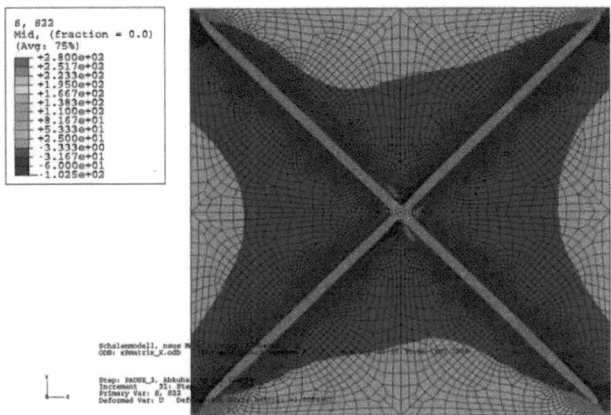

Figure 4.43: Fringe Plot of residual stresses σ_{yy} in MPa situated in the plate's mid-plane, heat input due to moved laser source, X-pattern geometry.

points, horizontally as well as vertically. The diameter of each laserpoint (two times the radius r_F) is chosen to be 9 mm. The sequence of pointwise laser irradiation (position of particular point) is denoted in Fig. 4.54. This sequence is chosen according to test simulations as well as experimental series to be the one resulting in the smallest residual deformations.

For the pointwise laser treatment laser parameters stay the same except the laser power P_L, which is reduced compared to the continuous laser treatment models shown before to 200 W resulting in a for the structure relevant power given in Eq. 4.3.

$$P^* = 0.86 \cdot P_L \cdot A_L = 0.86 \cdot 200W \cdot 0.3 = 51.6W \tag{4.3}$$

Duration of irradiation for each point is chosen to be 1.8 s. As a result, with the choice of the decreased laser power in combination with the chosen duration of irradiation, temperatures on the top-surface of the plates similar to the simulations discribed so far are achieved. Figure 4.55 shows the temperature fringes during introduction of laser point no. 14. Differences in the maximally reached temperatures between plate top-surface and plate bottom-surface, that are present in the plate,

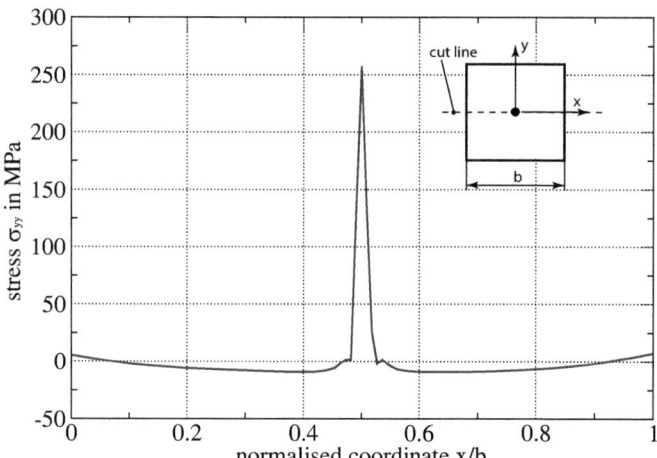

Figure 4.44: σ_{yy}-stress distributions (in MPa), along $y \equiv 0$, after heat treatment with moving laser source (X-pattern geometry). Stresses are situated in in the plate's mid-plane.

Chapter 4 Laser Treatment of Plate-Structures

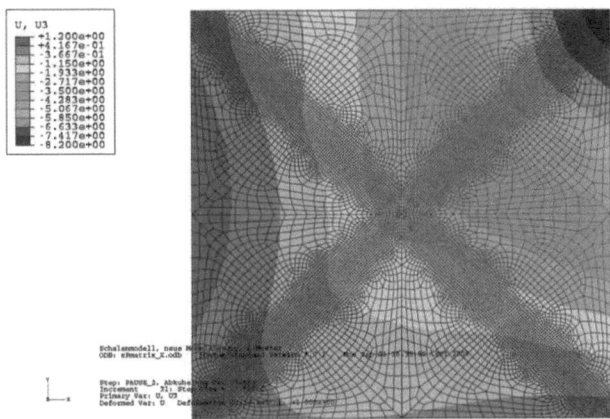

Figure 4.45: Fringe Plot of residual out-of-plane displacements U3 in mm. Heat treatment with moving laser source, X-pattern geometry.

are about 33 °C. This value is about twice the value observed in the simulations with a moving laser source.

Pointwise laser treatment is simulated in such a way that for each point the laser irradiation is performed (position of the point according to the irradiation sequence given above) and afterwards a cooling down to surrounding temperature is simulated. Separating each pointwise heat input completely and choosing the given sequence of the points should decrease the resulting thermally induced residual distortions after the complete treatment. After completed cooling down the calculated residual stress component σ_{yy} is shown as fringe plot in Fig. 4.56. Noteworthy is that in each point a very localised stress peak appears, as shown in Fig. 4.57.

Thermally induced distortions are rather small for the pointwise treated plate type 25P and residual distortions after complete cooling down are shown in Fig. 4.58. Additionally, as it can be seen in Fig.4.59, residual deformations are substantially smaller than those present in simulations with continuous laser treatment. Even with a deformation scale of 10, the isometric view in Fig. 4.60 shows a rather flat

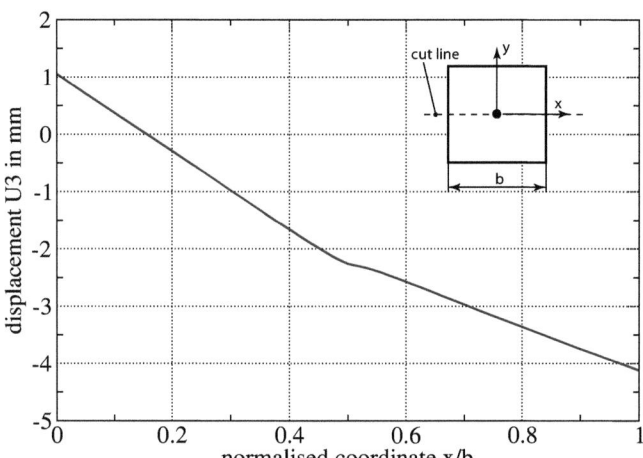

Figure 4.46: Residual out-of-plane displacements U3 in mm, along $y \equiv 0$. Heat treatment with moving laser source, X-pattern geometry.

Chapter 4 Laser Treatment of Plate-Structures

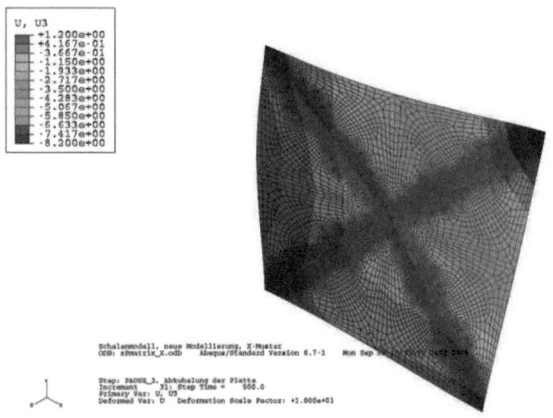

Figure 4.47: Fringe Plot of residual displacements U3 in mm under isometric view, deformation scale 10. Heat treatment with moving laser source, X-pattern geometry.

plate.

4.4 Evaluation of Laser Treatments with Respect to Buckling Resistance

In the following evaluation of the differently laser treated plate-structures the critical load concerning buckling (constant line load applied at the upper and lower plate edges in direction y and -y, respectively, resulting in a force controlled type of analysis) are investigated in linear buckling analyses. In addition to buckling analyses, the fundamental eigenfrequencies are calculated by a linear frequency analysis in which the external load is stepwise increased. This stepwise load increase allows to investigate the frequency development under increased loading, according to the kinetic stability criterion [5], as it is shown in forthcoming Chapters.

Chapter 4 Laser Treatment of Plate-Structures

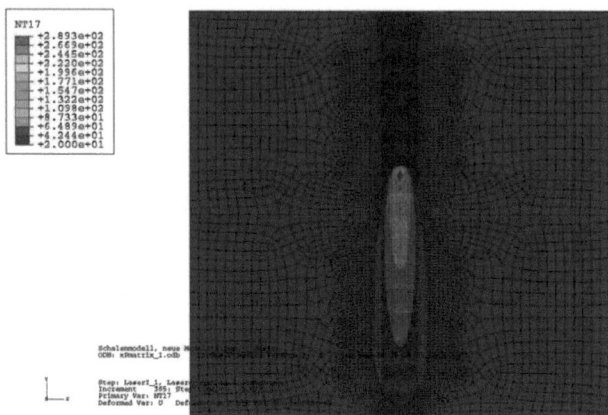

Figure 4.48: Fringe plot of temperature field in °C at the plate's top-surface, heat input due to moved laser source, I-pattern geometry.

4.4.1 Evaluation according to Buckling Load

The respective buckling load for each laser treated plate-structure, gained from linear buckling analysis, is given in Tab. 4.1. Boundary conditions used for buckling analyses correspond to the undeformed plate edge coordinate state. Therefore edges of treated plates are pulled back to this state with respect to their direction U3 (z-direction). Contrary to this, buckling load values which are given in Tab. 4.3 are obtained in analyses where the linear buckling evaluation is performed while boundary conditions are present on the deformed structure. At this point it has to be mentioned, that due to the thermally induced residual deformations a stress problem is present rather than a stability problem. Also because of this fact, results obtained by the linear buckling analysis can only be taken as estimates, and more detailed evaluations are gained by nonlinear force-displacement analyses.

For the plates treated by X - as well as circle - laser track geometries during pull back of the plate's edges into z-direction, to obtain the boundary conditions (case investigated in Tab. 4.1) and also during the application of the preload for the ac-

Chapter 4 Laser Treatment of Plate-Structures

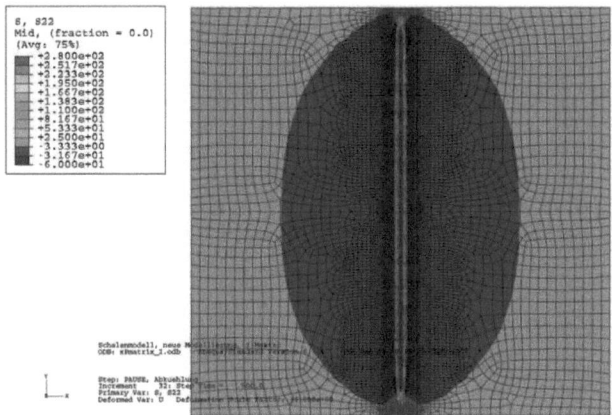

Figure 4.49: Fringe Plot of residual stresses σ_{yy} in MPa in the plate's mid-plane, heat input due to moved laser source, I-pattern geometry.

companying eigenfrequency analysis, local plastic deformations in the vicinity of the laser tracks appear. Because of that reason the conducted linear buckling analyses performed by ABAQUS FEM [1] programme are not directly suitable.

Application of Boundary Conditions on the Undeformed Edge Geometry, Pullback

In Fig. 4.61 the load displacement behaviour of plate-structures subjected to different laser treatment (boundary conditions after an edge 'pullback' in z-direction are obtained) is compared to that one for untreated plate-structures. For the untreated plate (BASE IMP) a small predeformation (according to the 1st buckle mode of a perfect plate, maximum translatorical out-of-plane displacement of $\frac{1}{1000}$ mm) is applied to act as imperfection enabling a reproducibility of the load-displacement behaviour of a nearly perfect plate within a nonlinear simulation.

83

Chapter 4 Laser Treatment of Plate-Structures

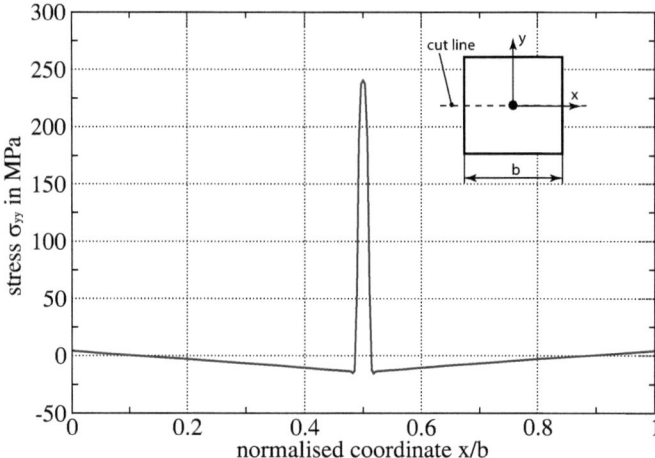

Figure 4.50: σ_{yy}-stress distribution (in MPa) in the plate's midplane along $y \equiv 0$ after heat treatment with moving laser source, I-pattern geometry.

Treatment	σ_{el}^* in [N/mm²]	Buck. Load F_q in N	Fund. Frequ. in Hz
untreated	5.49	1153.6	47.3
X-geo.	10.24	2150.4	65.4
I-geo.	7.82	1642.2	56.4
circle-geo.	9.81	2061.9	62.7
25P pattern	7.56	1586.7	55.7

Table 4.1: Critical compressive-stress σ_{el}^* for elastic buckling and the fundamental frequency for plate-structures subjected to different thermal treatment, boundary conditions are applied on the undeformed edges (pullback).

Chapter 4 Laser Treatment of Plate-Structures

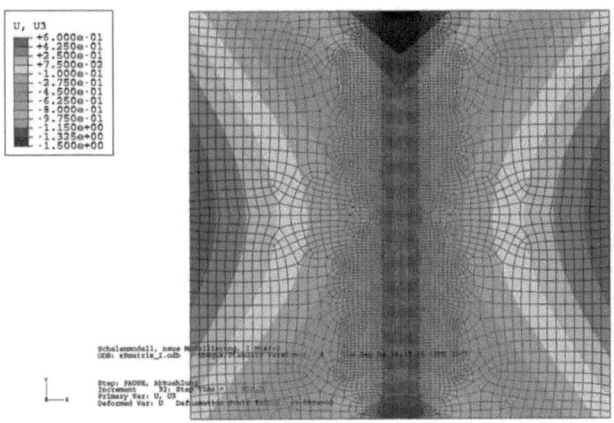

Figure 4.51: Fringe Plot of residual displacements U3 in mm. Heat treatment with moving laser source, I-pattern geometry.

Application of Southwell's Method

In the following considerations Southwell's method, as described in Chapter 2.3, is not only used for experiments but also for evaluating the computationally determined structural behaviour. Similar to buckling experiments the laser treated structures, considered in the computational buckling analyses, contain unavoidable geometrical imperfections and the following questions should be answered: which critical load could be assigned to the structure containing residual stresses but no residual distortions, i.e., no geometrical imperfections.

For a detailed look at the load-displacement behaviour of the laser treated plate-structures see Fig. 4.62. In this figure not the absolute displacement in z-direction of the plate's midpoint is shown, but the difference $\Delta_{(F)}$ for which Eq. 4.4 holds. Hereby $\delta^L_{(F)}$ stands for the absolute, load-dependent midpoint displacement value. The residual displacement of the midpoint of the unloaded plate is given by δ_0. This definition allows for an arrangement of the respective 'shifted' load-displacement

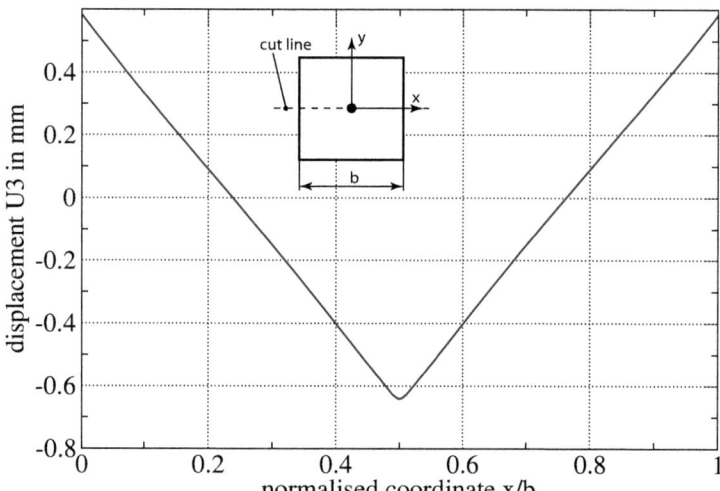

Figure 4.52: Residual out-of-plane displacements U3 in mm along $y \equiv 0$. Heat treatment with moving laser source, I-pattern geometry.

Chapter 4 Laser Treatment of Plate-Structures

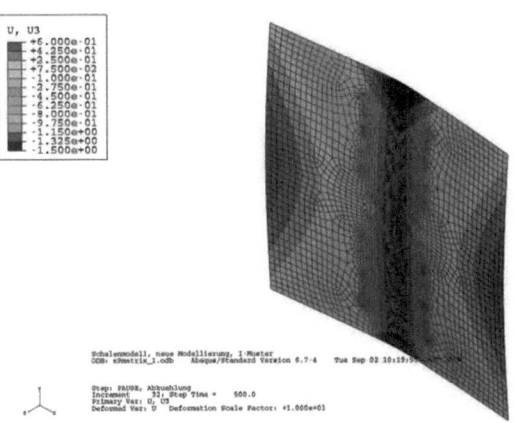

Figure 4.53: Fringe Plot of residual out-of-plane displacements U3 in mm, deformation scale 10. Heat treatment with moving laser source, I-pattern geometry.

Figure 4.54: Sequence of point position's for laser treated plate type 25P.

Chapter 4 Laser Treatment of Plate-Structures

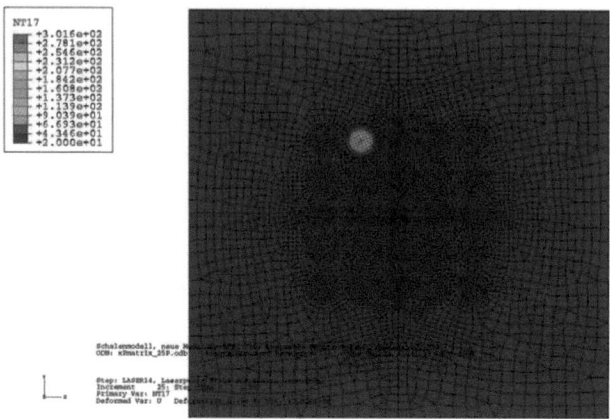

Figure 4.55: Fringe plot of temperature field in °C at the plate's top-surface, heat input due to pointwise laser treatment, 25P-pattern geometry.

curves, as shown in Fig. 4.62, obtaining a common origin in the diagram.

$$\Delta_{(F)} = \delta^L_{(F)} - \delta_0 \tag{4.4}$$

Applying Southwell's method one obtains a quasi-critical stress σ^*_q and a quasi buckling-load F_{qq}, for the treated structures in terms of an estimation of the critical load of the perfect structure, i.e., the bifurcation load, as given in Tab. 4.2.

Application of Boundary Conditions to the deformed Edges of Plate Structures.

Significant differences exist in the load-displacement behaviour of laser treated plate-structures (boundary conditions applied to deformed structures) compared with those of untreated plates, see Fig. 4.63. Due to the partially severe residual deformations on the laser treated plates, some load-displacement curves start with a significant horizontal offset on the x-axis (representing residual displacement in

Chapter 4 Laser Treatment of Plate-Structures

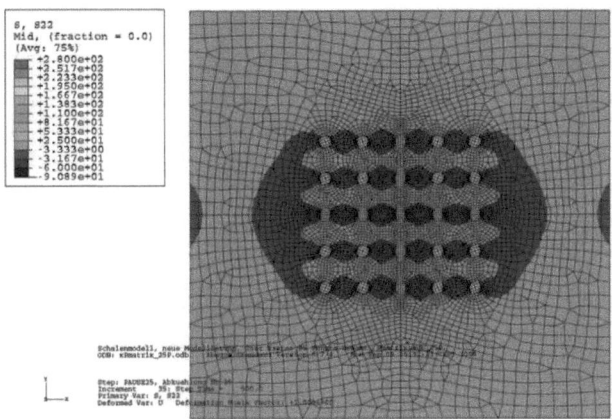

Figure 4.56: Fringe Plot of residual stresses σ_{yy} in MPa situated in the plate's mid-plane, heat input due to pointwise laser treatment, 25P-pattern geometry.

Treatment	σ_q^* in N/mm²	Buck. Load F_{qq} in N	Diff. to FEM in %
untreated-reference	5.49	1153.6	-
X-geo.	13.15	2762	28 ↑
I-geo.	8.37	1758	7 ↑
circle-geo.	9.7	2037	1 ↓
25P pattern	8.13	1707	8 ↑

Table 4.2: Quasi-critical compressive-stress σ_q^* and quasi-buckling load F_{qq} as well as difference to FEM buckling analyses in per cent. Results are for the laser treated structures, featuring a 'pullback' of the boundary conditions, under application of Southwell's method for 'shifted' load-displacement diagrams.

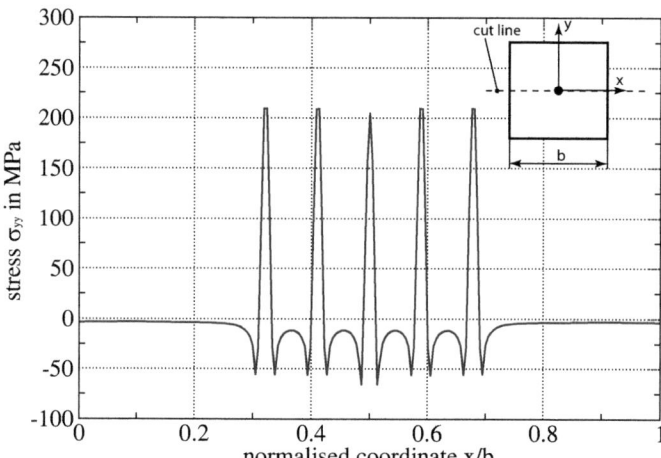

Figure 4.57: Residual σ_{yy}-stress distribution in MPa along $y \equiv 0$ in the plate's mid-plane after discontinuous, pointwise laser treatment (25P-pattern geometry).

Chapter 4 Laser Treatment of Plate-Structures

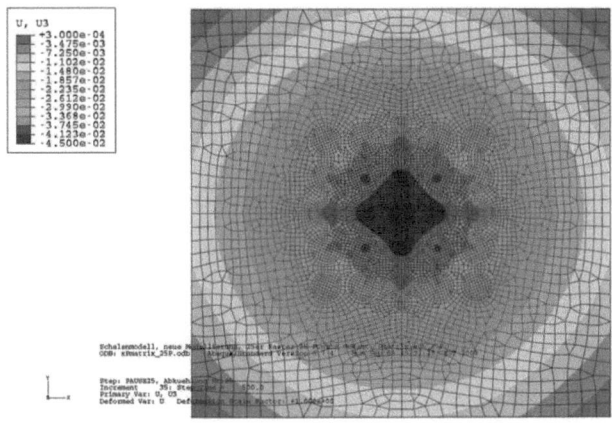

Figure 4.58: Fringe Plot of residual out-of-plane displacements U3 in mm. Heat input due to pointwise laser treatment, 25P-pattern geometry.

z-direction after laser treatment). Also for the current investigations a local plastic yielding occurs for the simulation model utilising a laser treatment with circle-geometry. Results for the present case concerning the elastic buckling load and the fundamental frequency are given in Tab. 4.3.

Detailed Comparison between Plate 25P and untreated Plate

A detailed comparison between the load-displacement behaviour of the untreated plate-structure and two cases of plate-structures treated with 25 laser points is shown in Fig. 4.64. Here, because of computational requirements, the untreated plate-structure contains a very small geometrical imperfection which is affine to the 1st-buckling-mode. For the laser treated plate in case 1 (plate 25P-PB), analysis boundary conditions are applied after pullback of the edges in z-direction. In case 2 (plate 25P), the laser treated plate features analysis boundary conditions which are applied onto the deformed edges. In both cases pointwise laser treatment shows a significant change in the load-displacement behaviour towards higher buckling

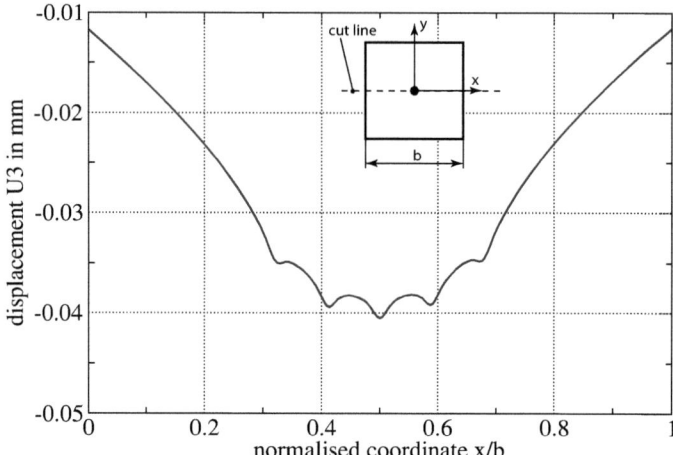

Figure 4.59: Residual out-of-plane displacements U3 in mm along $y \equiv 0$. Heat input due to pointwise laser treatment, 25P-pattern geometry.

Treatment	σ_{el}^* in N/mm²	Buck. Load F_q in N	Fund. Frequ. in Hz
untreated	5.49	1153.6	47.3
X-geo.	6.96	1461.6	52.8
I-geo.	5.88	1234.8	52.2
circle-geo.	8.15	1710.0	57.3
25P pattern	7.56	1587.6	55.6

Table 4.3: Critical compressive-stress σ_{el}^* for elastic buckling and the fundamental frequency for plate-structures subjected to different laser treatments; boundary conditions are applied to the deformed edges.

Chapter 4 Laser Treatment of Plate-Structures

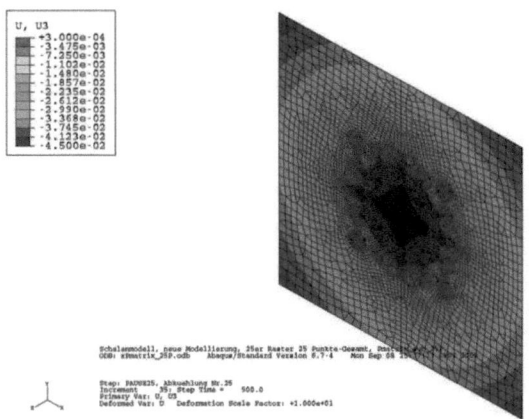

Figure 4.60: Fringe plot of residual out-of-plane displacements U3 in mm, deformation scale 10. Heat input due to pointwise laser treatment, 25P-pattern geometry.

resistance compared to the untreated plate-structure.

In the comparison of the two laser treated cases, only slight differences in their resulting load displacement behaviour are present. These results are due to only slight thermally induced residual deformations existing on the laser point treated plate-structures. The depicted stress problems are very close to results expected for stability problems. Thus, also by applying linear buckling analysis good estimations for critical loads can be expected.

4.4.2 Evaluation according to the Fundamental Frequency

Especially the 'square of the fundamental frequency' is a well suited indicator for the occurence of instabilities at perfect plate-structures. This is because under compressive load increase up to the critical load this value decreases linearly down to zero. For an investigation of this behaviour accompanying eigenfrequency analyses are performed. Starting at the unloaded state, with increasing the load stepwise, an

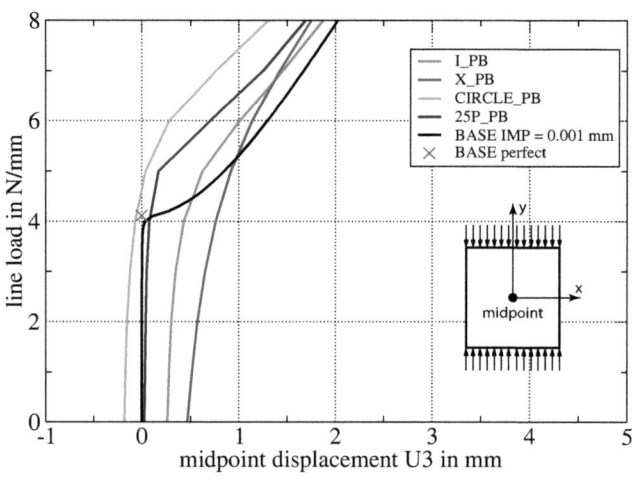

Figure 4.61: Load-displacement behaviour of laser treated and untreated plate-structures. Boundary conditions are in each case applied on the undeformed edges (edge-'pullback' denoted by acronym 'PB' in the respective legend).

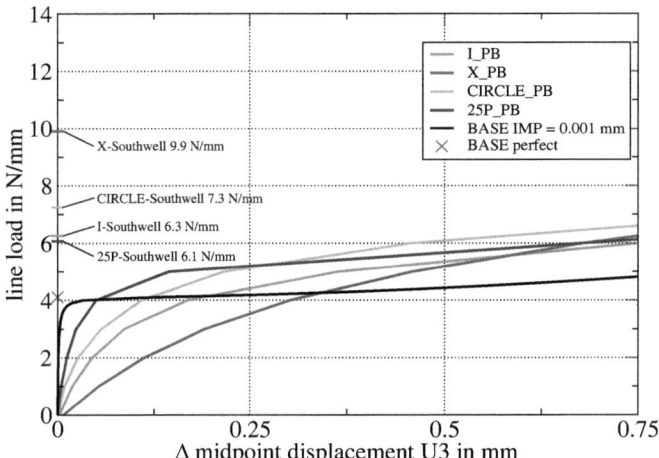

Figure 4.62: Load-displacement behaviour of laser treated and untreated plate-structures. The out-of-plane displacements in z-direction are in the current case Δ-values, according to Eq. 4.4. Boundary conditions are in each case applied on the undeformed edges (edge-'pullback' denoted by acronym 'PB' in the respective legend).

Chapter 4 Laser Treatment of Plate-Structures

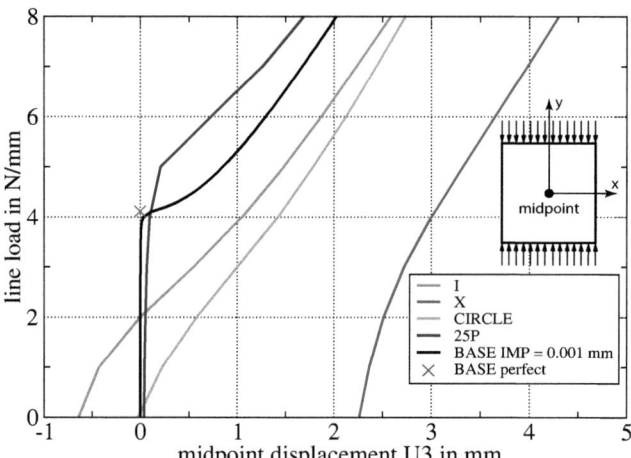

Figure 4.63: Comparison of laser treated and untreated plate-structures in their load-displacement behaviour, boundary conditions applied to the deformed edges.

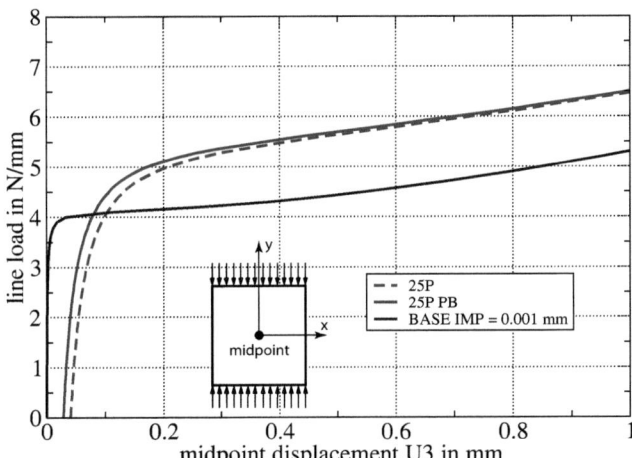

Figure 4.64: Load-displacement behaviour of the untreated base plate with included imperfection (1st buckling mode, normalised to maximum transversal displacement of 1 mm, of a perfect, untreated plate scaled with factor $\frac{1}{1000}$) in comparison to two cases of 25P-type laser treated structures. Case 1) boundary conditions applied to structure with edges after 'pullback', case 2): boundary conditions applied to the deformed structure.

Chapter 4 Laser Treatment of Plate-Structures

eigenfrequency analysis is conducted after each step. As load increase the raise of the line-load applied onto the plate's edges is meant, see Fig. 4.35 case b).

For imperfect plate-structures, at the beginning of the load increase a decrease of the fundamental frequency is noticeable, as expected. However, depending on the degree of imperfection resulting from different laser treatments, further increase of the load results also in an increase of the fundamental frequency. Therefore, an initial stiffness reduction is followed by a stiffness gain due to progressive deformations. For imperfect structures, like in the case at hand, this behaviour is an indicator that the stability problem merges into a stress problem.

Under usage of the formalism depicted in Chapter 2.4, the analytical solution for the eigenfrequency of an untreated perfect plate with geometrical and material parameters used here results in a value of 47.3 Hz for the fundamental frequency also according to a result obtained by the FE method.

Application of Boundary Conditions to Original Plate Edge Configuration - 'Pullback'

The fundamental frequency during a stepwise increase of the line-load, acting on the plate-structure's edges, is shown in Fig. 4.65 and likewise for the square of the fundamental frequency in Fig. 4.66. For all simulation results shown here, boundary conditions are applied on undeformed edges ('pullback').

Application of Boundary Conditions to deformed Edges

Results for the fundamental frequency development for boundary conditions on the deformed edges of the respective structure are shown in Fig. 4.67. Results for the development of the square of the fundamental frequency are depicted in Fig. 4.68.

A drop in the eigenfrequencies for the plate-structures, possessing severe pre-deformations, is noticeable in comparison to the plate-strutures analysed after an edge 'pullback'. Hence, applying a 'pullback' leads to a stiffening in terms of stress-rearrangements. More detailed investigations of consequences due to an application of different boundary condition on structures are presented in Chapter 4.5.

Chapter 4 Laser Treatment of Plate-Structures

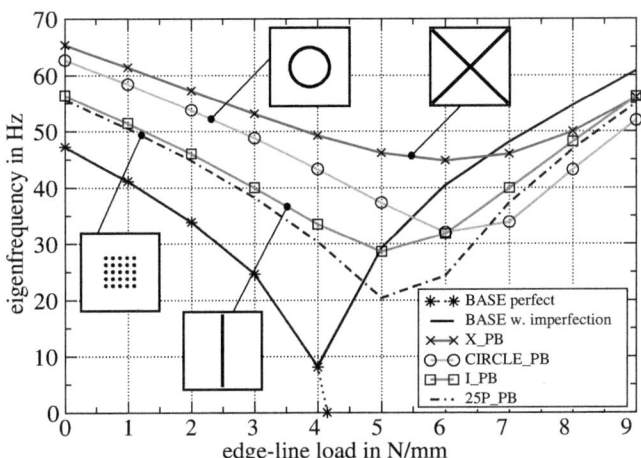

Figure 4.65: Fundamental frequency over the edge line-load for plate-structures subjected to different laser treatments, boundary conditions applied to undeformed edges ('pullback').

Chapter 4 Laser Treatment of Plate-Structures

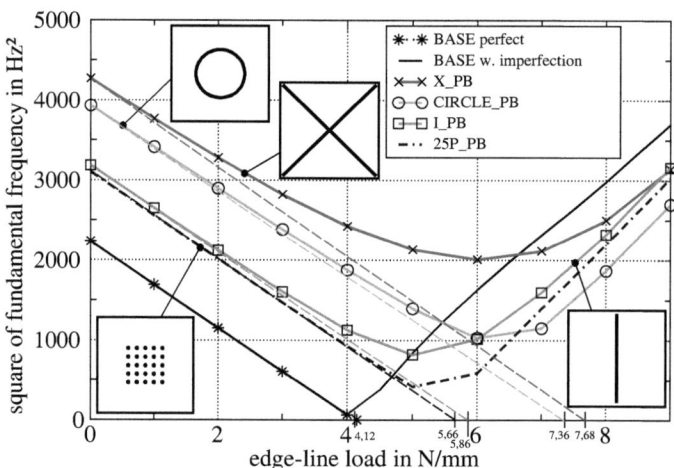

Figure 4.66: Square of the fundamental frequency over the edge line-load for plate-structures subjected to different laser treatments, boundary conditions applied to undeformed edges ('pullback').

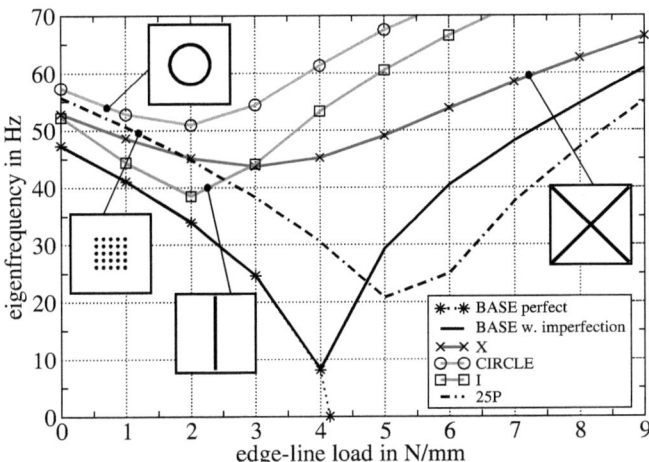

Figure 4.67: Fundamental frequency over the edge line-load in for plate-structures subjected to different laser treatments, boundary conditions applied on the deformed edges.

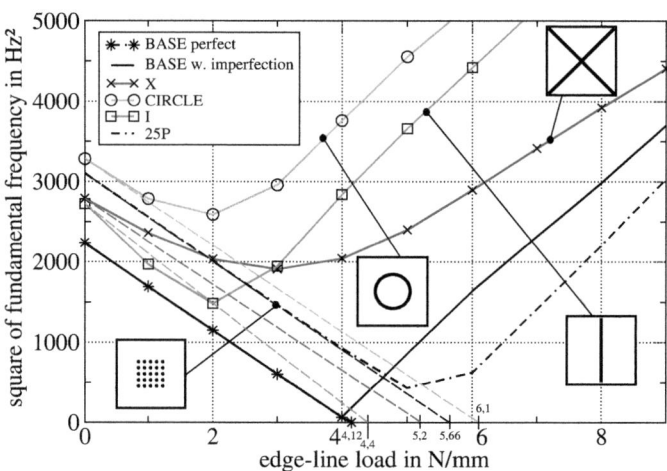

Figure 4.68: Square of the fundamental frequency over the edge line-load for plate-structures subjected to different laser treatments, boundary conditions applied on the deformed edges.

Material parameter	Value
Elastic modulus	$212.000 \, \text{MPa}$
Poisson's ratio	0.3
density	$7.85 \, \frac{\text{kg}}{\text{dm}^3}$

Table 4.4: Mild steel material properties applied for the beam stiffeners with quadratic cross section.

4.5 Analyses of Stiffened Plates with Laser Treatment

Additional investigations are conducted to study the influence of laser treatment of already stiffened plate-structures or plate-structures which are stiffened after treatement. In terms of a proof of concept, analyses of simple stuctures are performed. Industrial applications often desire more complicated configurations. For example, classical aircraft fuselages or railcar bodies often incorporate plate-structures stiffened by horizontal and vertical stringers. For this purpose investigations in the work at hand are extended to plate-structures combined with beam structures acting as stiffeners along the plate edges. Furthermore, in these investigations it is distinguished between weak- and strong laser treatments.

The investigated quadratic plate fields, acting as skins or shear webs, have an edge length $a = 280 \, \text{mm}$, as shown in Fig. 4.71, while exhibiting a thickness of $0.75 \, \text{mm}$. All plates have the above described material properties for the 'DC01'-steel, considering temperature dependent non-linear behaviour. In simulations where stiffeners are used, these feature a quadratic cross section with an edge length of $20 \, \text{mm}$. For the stiffeners simply linear elastic material behaviour is assumed, see Tab. 4.4.

Following cases are investigated:

- Case 1, reference plates, unstiffened and stiffened plate without laser treatment
- Case 2, unstiffened plates subjected to laser treatment
- Case 3, laser treatment of unstiffened plates, beam stiffeners are applied after completed laser treatment

Chapter 4 Laser Treatment of Plate-Structures

- Case 4, laser treatment of unstiffened plates, out-of-plane deformations on the edges are reset to zero, beam stiffeners are applied thereafter

- Case 5, laser treatment of already stiffenend plates

Laser treatment is simulated by introducing a pattern of 25 laser dots, all at the same time. Thereby the laser acts for a time span of 1.8 s. The resulting temperature distribution is shown in Fig. 4.69 for the weak laser treatment and in Fig. 4.70 for the strong laser treatment. In all cases where laser treatment is applied, weak laser treatment introduces a theoretical value of the laser power of $P_L = 200$ W and strong laser treatment a theoretical laser power value of $P_L = 800$ W. The distribution of the power density in each laser dot is of Gaussian type. For the weak laser treatment a focus radius of $r_F = 4.5$ mm is used in contrast to the simulation of the strong laser treatment of the structures, where $r_F = 9.0$ mm is used. For both cases the laser dot itself is cut at 'r_F' which implies that no laser power is introduced into the structure for radial distances $r > r_F$ (with reference to the origin of each investigated laser dot). As it has already been shown, a cutted Gaussian distribution and the same absorption coefficient on the plate's top-surface position $A_L = 0.3$ are used. Hence, the resulting power input into the structures is less than the theoretical values given above (51.6 W for the weak laser treatment and 206.4 W for the strong laser treatment). After the laser treatment, each treated structure is cooled-down to room temperature, 20 °C, by applying a cooling down time of 500 s under free convection at air.

During laser treatment boundary conditions, as shown in Fig. 4.71 a), are set. Buckling- and eigenfrequency analyses are conducted under application of the boundary conditions shown in Fig. 4.71 b).

The strong laser treatment of unstiffened plates results in drastic local- as well as global deformations. On the one hand, these deformations require the application of boundary conditions to the undeformed structure in such a way, that only slight deformations on the laser treated structure remain. On the other hand, the boundary conditions should not impose additional unwanted constraints and, therfore, resulting stresses in the respective structure. Boundary conditions applied while laser treatment, as shown in Fig. 4.71 a), represent such a trade-off. Precisely these boundary conditions not only constrain rigid body movements of the structure but also constrain the 'normal out-of-plane' movement of the north-east (NE) corner of

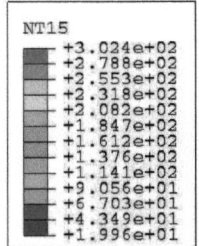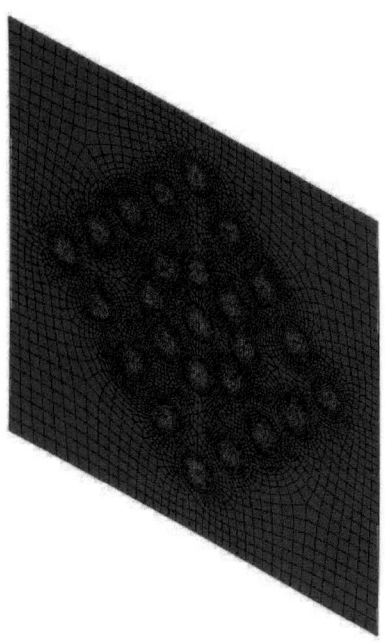

Figure 4.69: Fringe plot of temperature field in °C during weak laser treatment of plate A, see for example Chapter 4.5.2, on the plate's top-surface. Laser active for $t = 1.8\,\text{s}$, deformation scale 1.0.

Chapter 4 Laser Treatment of Plate-Structures

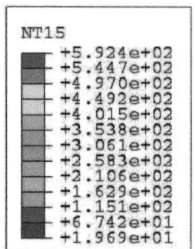

Figure 4.70: Fringe plot of temperature field in °C during strong laser treatment of plate AS, see for example Chapter 4.5.2, on the plate's top-surface. Laser active for $t = 1.8\,\text{s}$, deformation scale 1.

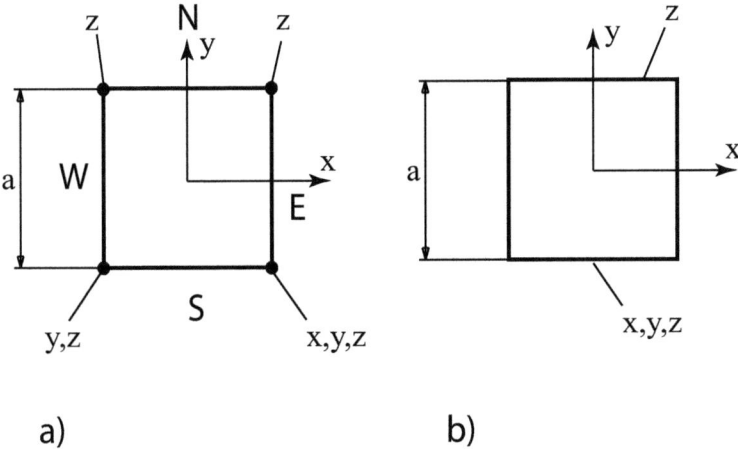

Figure 4.71: Boundary conditions present for investigations on 'stiffened' plates. a) boundary conditions during laser treatment, b) boundary conditions set for buckling- and eigenvalue analysis.

Chapter 4 Laser Treatment of Plate-Structures

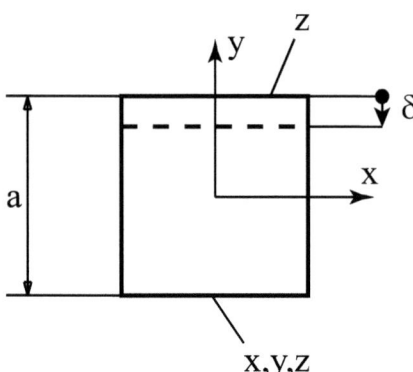

Figure 4.72: Direction δ of preset displacement during the buckling- analysis.

the plate. Thus, an out-of-plane deformation of the NE corner is reduced, but one cannot speak of global residual stresses anymore. Rather 'a sort' of residual stresses in the plane of the plate could be stated. As already described in Chapter 4.2.4, of particular interest are the normal stresses in the plate's mid-plane, in the direction of the external load (σ_{yy}-stresses). Additionaly, each setting of different boundary conditions as well as each activation/deactivation of stiffeners is treated in one separate analysis step. For the boundary conditions a special 'velocity' type is used to apply them on the deformed structure and fullfill the before statet necessity of not influencing the already existing stress state.

For the buckling analysis, a reference displacement δ in y-direction acts as 'reference load', see Fig. 4.72. Resulting λ values of the analyses have to be scaled with the reference displacement δ to obtain the critical displacement value.

All investigated plates, also the untreated types, include an imperfection which corresponds to the first buckling mode of the untreated perfect plate, having an imperfection amplitude of 0.1 mm. This predeformation should ensure the ability of the simulation to follow the non-trivial equilibrium path also for perfect structures (e.g., the untreated plates).

Chapter 4 Laser Treatment of Plate-Structures

4.5.1 Case 1, untreated reference plates

investigated structures:
PA_U; untreated plate without stiffeners
PB_U; untreated plate with stiffeners

Both plate types, the untreated unstiffenend plate as well as the untreated stiffened plate are investigated to obtain reference solutions.

For the plate PA_U its behaviour corresponds to a 'two sided-free' plate where boundary conditions act only on the top edge (under displacement load) and the bottom edge. Critical displacement as well as fundamental frequency are denoted in Tab. 4.5. Resulting modes from both the buckling analysis, shown in Fig. 4.73, as well as the eigenfrequency analysis, Fig. 4.74, show virtually the same picture. In order to avoid confusion it should be mentioned that for untreated plate-structures the same FE-mesh was used as for the laser treated ones.

Regarding the structure PB_U, the beams contribute by a stiffening effect which results in a ca. 9.4-times higher critical displacement as well as an about 3.7-times higher value of the fundamental frequency compared to structure PA_U. Detailed values can be found in Tab. 4.5. As expected, the first buckling mode, see Fig. 4.75, and also the fundamental vibration mode, displayed in Fig. 4.76, show a behaviour which is known from plates clamped along all edges. Again both modes show a very similar picture.

4.5.2 Case 2, unstiffened plates under laser treatment

investigated structures:
PA; unstiffened plate subjected to weak laser treatment
PAS; unstiffened plate subjected to strong laser treatment

In case 2 the original plate PA_U is treated by weak as well as strong laser irradiation. A pattern consisting of 25 laser points is simultaneously applied with the specifications given above (see Chapter 4.5). As statet before, during laser treatment drastic deformations appear. For the weak laser treatment a maximum deformation of $U_{mag}^{max} = 3.7\,\text{mm}$ arises in the plate's midpoint during the maximum occuring

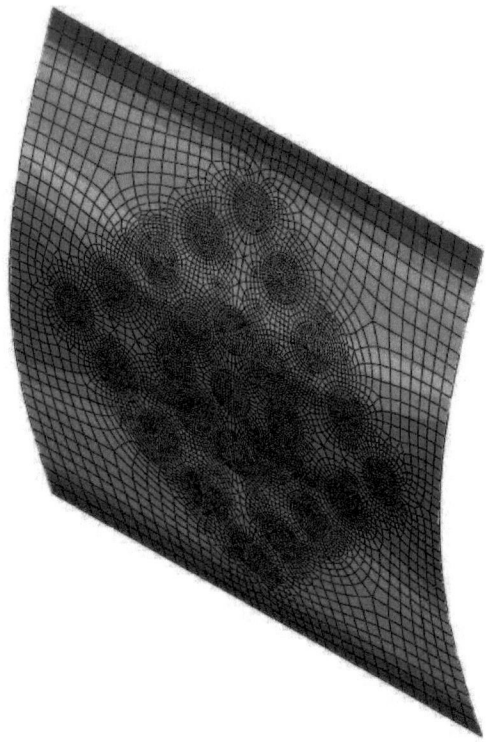

Figure 4.73: First buckling mode of plate PA_U (untreated plate without stiffeners). The FE-mesh used is the same as it serves for treated plates.

Chapter 4 Laser Treatment of Plate-Structures

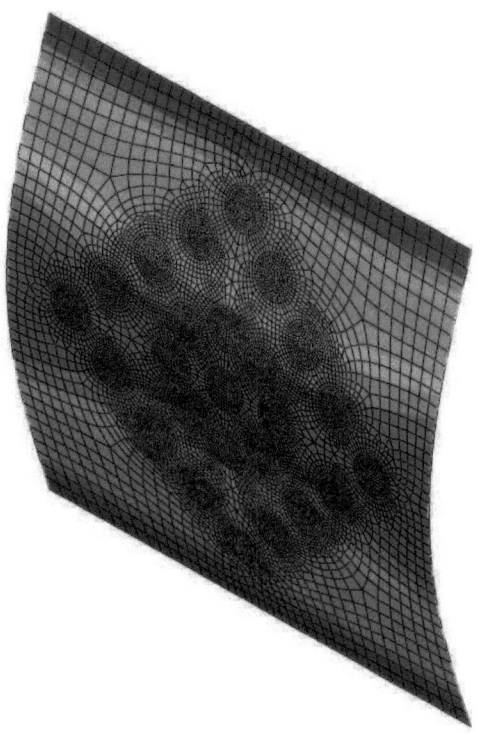

Figure 4.74: Mode of fundamental frequency of plate PA_U (untreated plate without stiffeners).

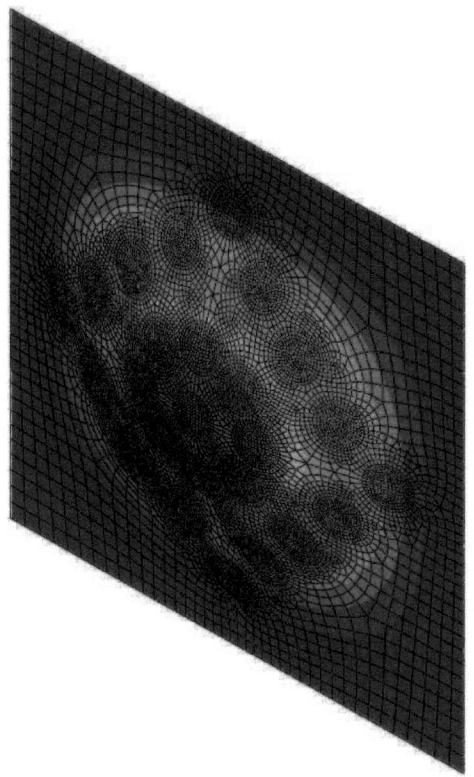

Figure 4.75: First buckling mode of plate PB_U (untreated plate with stiffeners).

Chapter 4 Laser Treatment of Plate-Structures

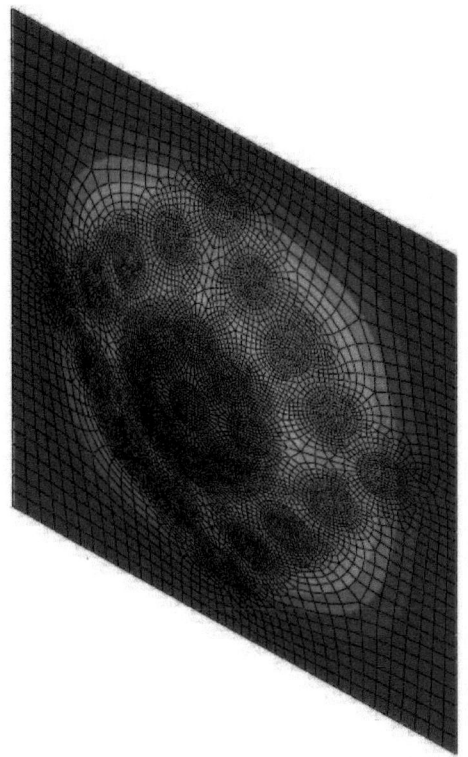

Figure 4.76: Fundamental vibration mode of plate PB_U (untreated plate with stiffeners).

Chapter 4 Laser Treatment of Plate-Structures

temperatures. After complete cooling down, remaining deformations are reduced to $U_{mag}^{max} = 0.27$ mm. During the strong laser treatment the deformations even reach maximum values up to $U_{mag}^{max} = 11.5$ mm and remaining values at room temperature are at $U_{mag}^{max} = 3.54$ mm. In Fig. 4.86 residual deformations along a horizontal cut through the structure's midpoint are shown for the investigated cases.

Regarding buckling resistance and fundamental frequency of PA and PAS, respectively, values are given in Tab. 4.5. Compared to plate PA_U, shown in Chapter 4.5.1, the first buckling mode and the mode of the fundamental frequency show nearly the same form, exemplarily the first buckling mode is shown in Fig. 4.77.

A different result arisis when investigating the first buckling mode of structure PAS, see Fig. 4.78. Firstly a completely different mode shape compared to the untreated structure PA_U is visible. Furthermore, the resulting value of critical displacement is about 174 times higher than the untreated structure version PA_U. Nevertheless, only a small gain in the fundamental frequency, also in comparison to structure PA_U, is noticeable. Detailed results concerning critical displacement and fundamental frequency are given in Tab. 4.5. A comparison between occuring residual stresses σ_{yy} along a horizontal path through the plate's midpoint after cooling down in structure PA and structure PAS is displayed in Fig. 4.79. The strong laser treatment increases both tensile stresses and compressive stresses in the vicinity of the point of laser treatment compared to the results obtained through weak laser treatment. The structure PAS, which is subjected to strong laser treatment, shows only a very small increase in the fundamental frequency but an extreme increase in the critical displacement. These results are obviously present due to massive deformations resulting from the laser treatment.

For the sake of completeness the variation of the longitudinal residual stresses, resulting from each of the two described types of laser treatment (weak and strong) along the line $y \equiv 0$ for different positions over the plate's thickness, is also investigated for plate PA as well as plate PAS. Using a total number of five section points (SP, ABAQUS notation) for investigations at hand with Simpson's integration method leads to Fig. 4.80, for plate PA. The positions of the section points SP along the plate's thickness are given in the figure legend. It can be seen that for weak laser treatment the distribution of the residual stresses, except for peak points, does not dramatically differ along the plate's thickness. For the strongly

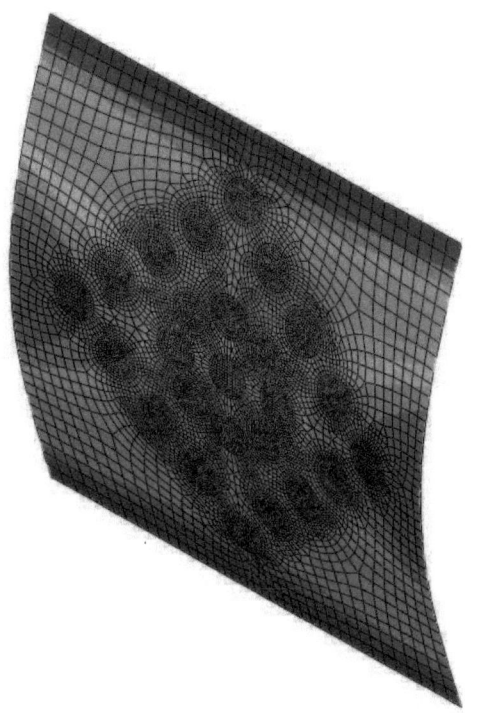

Figure 4.77: First buckling mode of plate PA (unstiffened plate subjected to weak laser treatment).

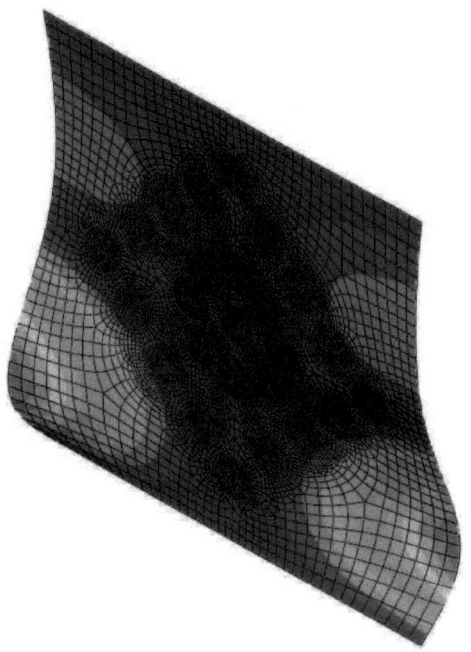

Figure 4.78: First buckling mode of plate *PAS* (unstiffened plate subjected to strong laser treatment).

Chapter 4 Laser Treatment of Plate-Structures

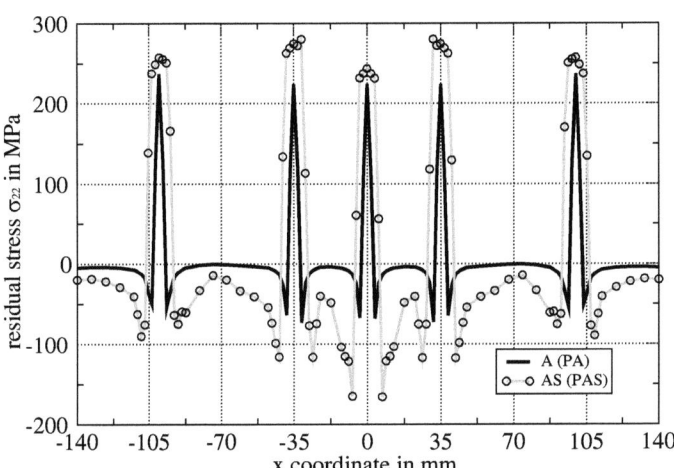

Figure 4.79: Longitudinal residual stresses σ_{yy} of laser treated plates A (PA) and AS (PAS).

Chapter 4 Laser Treatment of Plate-Structures

laser treated plate the σ_{yy} stress distributions in different section points are shown in Fig. 4.81. There, significant variations along the plate's thickness can be found.

4.5.3 Case 3, laser treatment of unstiffened plates, beam stiffeners are mounted after completed laser treatment

investigated structures:

A; weak laser treatment of unstiffened plate, stiffeners are applied after laser treatment

AS; strong laser treatment of unstiffened plate, stiffeners are mounted after laser treatment

From FEM-modelling point of view originally deactivated beam elements are reactivated in an own analysis step resulting in a stiffened structure where no additional constraints are imposed. As a consequence the subsequently stiffened plates A and AS correspond, regarding to their residual deformations, to the in case 2) investigated plates PA and PAS. The resulting deformation distributions can be seen in Fig. 4.86. Regarding enhancement in both buckling resistance and fundamental frequency weak laser treatment contributes in a way that structure A surpasses the untreated stiffend plate B by ca. 12% w.r.t. critical displacement as well as about 7% w.r.t. fundamental frequency. As already observed in case 2) (structure PAS) strong laser treatment of structure AS leads to a dramatic increase in the critical displacement. A heavy increase of the fundamental frequency, in comparison to untreated structures, appears. For comparison complete data is given in Tab. 4.5.

The mode shapes for the weakly laser treated plate (both first buckle mode and fundamental vibration mode) show nearly the same picture and are qualitatively directly comparable to structure PB_U (which is shown in Fig. 4.75 as well as Fig. 4.76). Contrary to this, for the structure AS, which is subjected to strong laser treatment and stiffening afterwards, a mode change both for the buckling mode and the first vibration mode sets in. In terms of σ_{yy} stress distributions the same characteristics as shown in Fig. 4.80 (structure PA) and Fig. 4.81 (structure PAS) are valid for structures A and AS, respectively. This results from the same thermal treatment, in each distinct case, and the uninfluencing mounting of stiffeners.

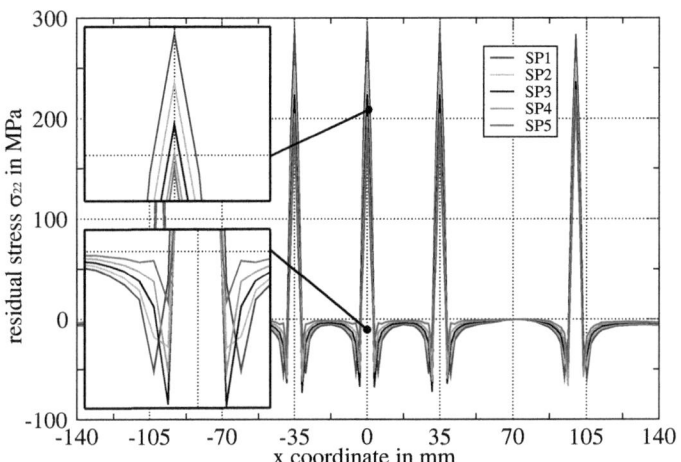

Figure 4.80: Longitudinal residual stresses σ_{yy} of weakly laser treated plates A (PA) after complete cooling down in different section points (SP). SP1 (-0.375 mm) plate's bottom-side (averting to laser beam), SP2 (-0.1875 mm), SP3 (0 mm) plate's mid fraction, SP4 (0.1875 mm), SP5 (0.375 mm) plate's top-side.

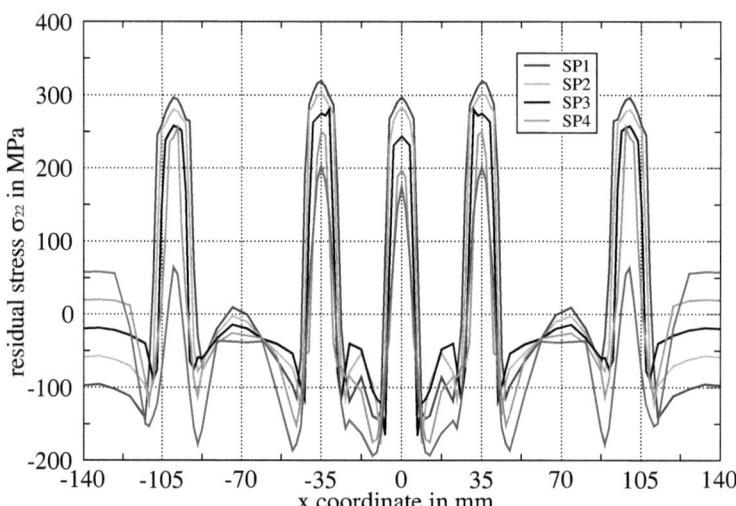

Figure 4.81: Longitudinal residual stresses σ_{yy} of strongly laser treated plates AS (PAS) after complete cool-down in different section points (SP). SP1 (-0.375 mm) plate's bottom-side (averting to laser beam), SP2 (-0.1875 mm), SP3 (0 mm) plate's mid fraction, SP4 (0.1875 mm), SP5 (0.375 mm) plate's top-side.

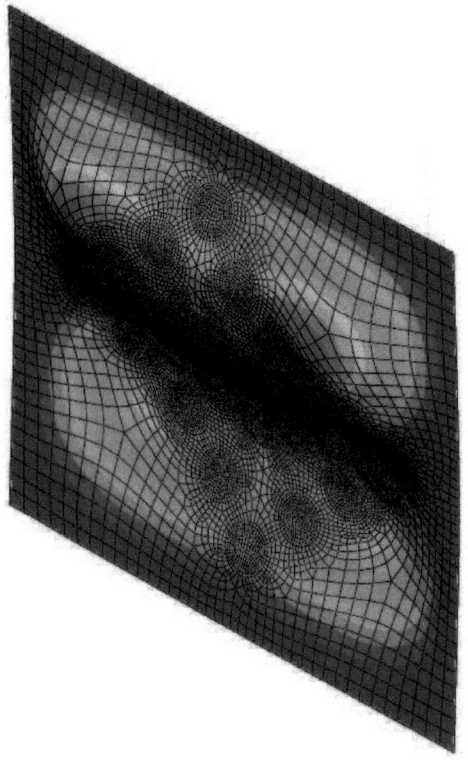

Figure 4.82: First buckling mode of plate AS (strong laser treatment of unstiffened plate, stiffeners are mounted after laser treatment).

Chapter 4 Laser Treatment of Plate-Structures

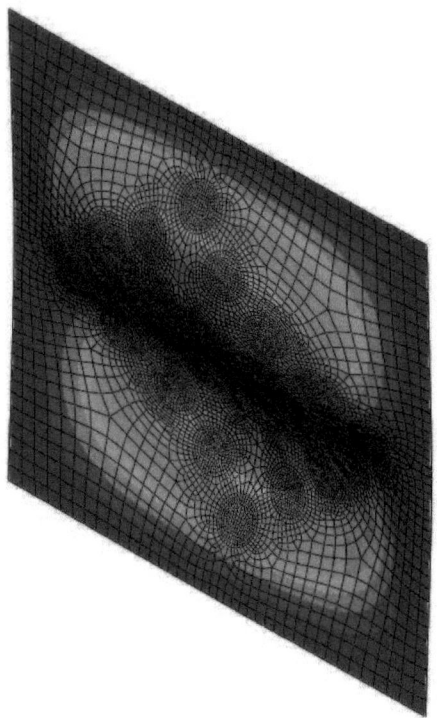

Figure 4.83: Fundamental vibration mode of plate *AS* (strong laser treatment of unstiffened plate, stiffeners are mounted after laser treatment).

Chapter 4 Laser Treatment of Plate-Structures

As already mentioned in case 2), expectations of enhancements due to residual stresses were not met because of massive deformations after thermal treatment. This fact also holds for structures investigated in the current case 3). Stiffeners at least provide a substantial increase in the fundamental frequency.

4.5.4 Case 4, laser treatment of unstiffened plates, out of plane deformations (direction z) on the edges are reset to zero, beam stiffeners are mounted thereafter

investigated structures:

A_0; weak laser treatment of unstiffened plate, out-of-plane deformations (z-direction) on the edges are reset to zero

AS_0; strong laser treatment of unstiffened plate, out-of-plane deformations (z-direction) on the edges are reset to zero

As a consequence of results obtained in case 3), which show poor comparability to untreated structures, investigations with boundary conditions under edge 'pullback' conditions are performed. In the case at hand, boundary conditions after laser treatment are rearranged on the already investigated structures A and AS (case 2). Hereby the structure's edges are pulled-back into the originally undeformed configuration in direction 3 (z-coordinate). Figure 4.71 shows the boundary conditions during the laser treatment, and in Fig. 4.84 the configuration of the pulled back boundary condition state is depicted. A direct comparison of the structures A_0 and AS_0 to the unstiffened ones, PA_U as well as PB_U, shows an to expectations consistent increase in both critical displacement and fundamental frequency; detailed results are given in Tab. 4.5. For instance, in a comparison between AS_0 and untreated structure PB_U one can see an increase of about 98% regarding to the critical displacment and over 40% increase in the fundamental frequency; all values are summarised in Tab. 4.5. These increased values can be seen as direct result from emerging residual stresses. To be precise one has to speak of stresses in the structure which is subjected to the given treatment- and boundary conditions, rather than of real residual stresses. Further, maximum residual displacements are beyond 0.5 mm which is marginal compared to the maximum deformations appearing in structure

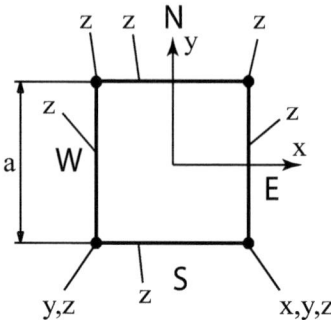

Figure 4.84: Boundary conditions in case of an edge-'pullback'.

AS (exceeding 3.5 mm), compare Fig. 4.86.

As stated above, a 'pullback' of the treated structures (to set back the structure's edges in z-direction) gives a more realistic configuration of the stiffened structure in a way that, for example, it can be added to an existing structure without the limitation of too large deformations.

Regarding the mode shapes of structure A_0, compared to structure A in case 3), no change in the mode shape, for both buckling and fundamental vibration mode, occurs. Contrary to this, AS_0 shows a change in the first buckling mode (in comparison to A_0) by an increase in the half-wave number, see Fig. 4.85 (first buckling mode of structure AS_0). The same change in the first buckling mode holds for structure AS, investigated in case 3). However, looking at the first vibration mode one cannot see a change compared to A_0. Whereas, there is a change compared to AS where, as stated above, also a change in the half-wave number of the fundamental vibration mode appears.

Chapter 4 Laser Treatment of Plate-Structures

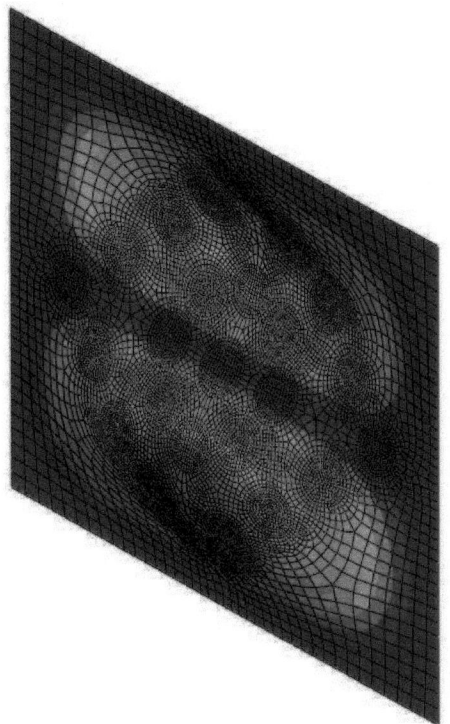

Figure 4.85: First buckling mode of plate AS_0 (strong laser treatment of unstiffened plate, out-of-plane deformations, in z-direction, on the edges are reset to zero).

4.5.5 Case 5, laser treatment of already stiffenend plates

investigated structures:
B; weak laser treatment of already stiffened plate
BS; strong laser treatment of already stiffened plate

Investigations carried out for these stiffened structures, which also feature weak and strong laser treatement, show again a consistent behaviour compared to the untreated structures. The highest increases appear for the strongly laser treated structure BS in both critical displacement (BS features 2.7 times the value of the untreated, stiffened structure PB_U) as well as in the fundamental frequency (a 70% increase for structure BS compared to structure PB_U). Details to all results are given in Tab. 4.5. Concerning mode shapes, the same applies as shown in case 4). The first buckling mode changes for the strongly laser treated structure while the fundamental vibration mode stays the same for both. Residual deformations of the respective structures are shown in Fig. 4.86. The deformations are in general very small, especially near to the structure's edges where they show some kind of horizontal tangent similar to clamped-type boundary conditions.

4.5.6 Conclusion for laser treatment of prestiffened plate-structures

To sum up the results obtained from investigations in cases 1) to 5) one can say that laser treatment of already stiffened structures, because of the artificial constraints, shows an increase in residual stresses in the structure at a smaller amount of overall deformations. This effects can even be increased if a more powerful laser treatment is applied. When there is no possibility to apply laser treatment to an already assembled structure it is also possible to apply laser treatment of the plate first, and subsequent assemble the structure. However, depending on the laser treatment, laser irradiation on unstiffened plates might imply massive residual deformations and such plates have to be reformed to fit to the assembly. In general a laser treatment of already stiffened structures is more efficient than the second possibility of divided laser treatment and assembly.

Chapter 4 Laser Treatment of Plate-Structures

plate-structure	Crit. displacement λ_1 in mm	Eigenfrequ. f_1 in Hz
Case 1		
PA_U	$1.71 \cdot 10^{-3}$	23
PB_U	$1.61 \cdot 10^{-2}$	86
Case 2		
PA	$2.09 \cdot 10^{-3}$	21.6
PAS	0.298	28.8
Case 3		
A	$1.92 \cdot 10^{-2}$	92
AS	0.17	189.4
Case 4		
A_0	$1.88 \cdot 10^{-2}$	91.6
AS_0	$3.181 \cdot 10^{-2}$	121.6
Case 5		
B	$2.08 \cdot 10^{-2}$	96.7
BS	$4.45 \cdot 10^{-2}$	146

Table 4.5: Results for the critical displacement and eigenfrequency for the investigated cases 1 to 5.

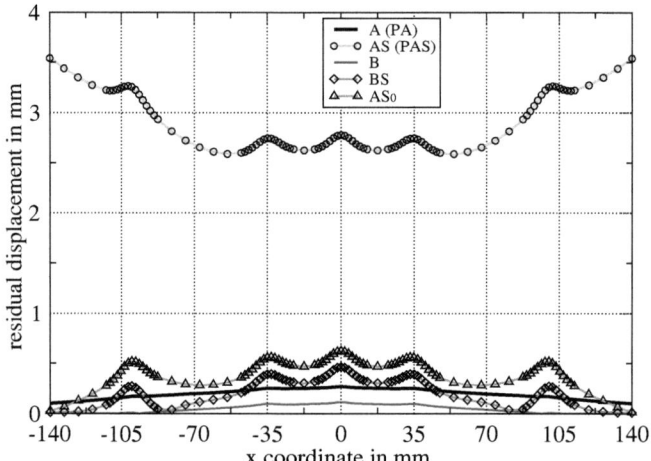

Figure 4.86: Residual displacements for laser treated structures along $y = 0$. Conditions of structures during laser treatment: A (PA)...unstiffened plate subjected to weak laser treatment, AS (PAS)...unstiffened plate subjected to strong laser treatment, B...weak laser treatment of already stiffened plate, BS...strong laser treatment of already stiffened plate. Pullback of translatoric boundary condition z-direction: AS_0...originally unstiffened plate subjected to strong laser treatment.

Chapter 4 Laser Treatment of Plate-Structures

Figure 4.87: Introduction of a continuous laser track into a plate-structure, I-geometry. (courtesy of G. Figala, LUT, Montanuniversitaet Leoben)

4.6 Comparison with Experimental Results of Laser Treated Plates

All experimental investigations were performed at the LUT in Leoben. Details can be found in the technical report [21].

4.6.1 Experimental Laser Treatment of Plates

For the purpose of laser treatment two different laser types are used. A Nd-YAG laser, which is mounted on a CNC - roboter arm, is used for plates subjected to continuous and intermitting laser treatment. Thereby, laser power between 515 W and 590 W (effecting the plate-structure's surface) is used. In Fig. 4.87 treatment of a plate by a Nd-YAG laser (continuous line treatment; I-geometry) is shown as an example.

In the objective case, for continuous Nd-YAG laser treatment quite heavy residual distortions appear. For the laser point treatment the residual distortions after

Chapter 4 Laser Treatment of Plate-Structures

Figure 4.88: Residual deformations of laser treated plate-structures. Left picture shows plate after laser treatment with circle-geometry. Right picture shows plate after 25 point laser treatment. (courtesy of G. Figala, LUT, Montanuniversitaet Leoben)

complete cooling down are still visible. However, they are substantially smaller. Figure 4.88 shows two plate-structures after laser treatment. In the left picture residual deformations on a plate-structure, after Nd-YAG laser treatment along a continuous laser pattern (circle geometry), are shown. In the right picture a plate-structure after laser treatment on a 25 point pattern is shown. In a direct comparison it can be seen that smaller residual deformations occur for the plate subjected to pointwise laser treatment.

Although different experimental measures are applied (e.g., reduction of laser power, additional edge support), residual distortions cannot be completely avoided. This is mainly due to the construction principle of the used Nd-YAG laser which posseses a distinct minimum feasible laser power one cannot underrun. Because of that reason, additional experiments under usage of a diode laser (with laser power reducible to 300 W) are conducted.

4.6.2 Test Equipment

For buckling experiments an universal testing machine ZWICK/Roell Z 250 is used. For this machine special frames were designed and produced to facilitate the re-

Chapter 4 Laser Treatment of Plate-Structures

Figure 4.89: Testing machine for plate buckling investigations.
(courtesy of G. Figala, LUT, Montanuniversitaet Leoben)

spective plate boundary conditions. In Fig. 4.89 the testing machine setup is shown when testing a plate treated by continuous laser irradiation along a circular path.

4.6.3 Comparing Simulation and Experiment

Experiments show two main difficulties for a clear comparison with results obtained from the FEM-simulation. One difficulty arises already from the test of untreated plates. In the experiments even these structures mostly show load-displacement behaviours which typically appear in simulations of plate-structures containing pre-deformations. This means, that even the untreated plates exhibit substantial imperfections due to the manufacturing process. Furthermore, boundary conditions, applied in the simulation, are hardly realised in the experimental setup.

Experimentally determined load-displacement curves of both untreated and laser treated plate-structures (plate type D10-light diode laser treatment, plate type V1-strong NdYAG laser treatment, both pointwise laser treatment in diamond shape) are shown in Fig. 4.90. Figure 4.91 shows a comparison between load-displacement

Chapter 4 Laser Treatment of Plate-Structures

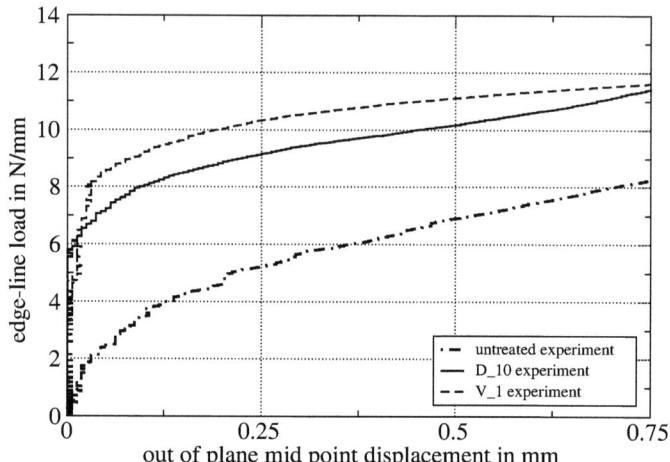

Figure 4.90: Experimentally obtained load-displacement curves for different plate-structures. (experimental data courtesy of G. Figala and B. Buchmayr, Lehrstuhl für Umformtechnik, LUT, Montanuniversitaet Leoben)

curves computed for plates with simply supported and clamped boundary conditions, with results from experiments under usage of boundary conditions intended to be simply supported. For this investigation of the influence of boundary conditions, the type D10 plate-structure, which was subjected to pointwise laser treatment by the diode laser, was used.

4.7 Discussion of obtained Results

The actual FEM models used to simulate the introduction of residual stresses into plate-structures by laser irradiation show multiple effects. In the case of continuous

Chapter 4 Laser Treatment of Plate-Structures

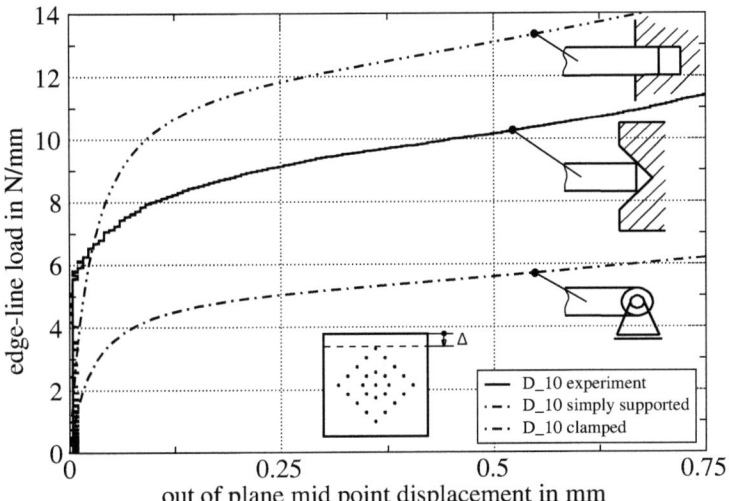

Figure 4.91: Boundary condition influence on the load-displacement behaviour of a plate-structure (simulation vs. experiment). (courtesy of G. Figala and B.Buchmayr, LUT, Montanuniversitaet Leoben)

Chapter 4 Laser Treatment of Plate-Structures

laser treatment the thermally induced strains and resulting stresses are - because of the one-sided laser irradiation - nonsymmetric w.r.t. the mid-plane and reach values leading to substantial overall deformations of the treated plate-structure. These deformations effect, under external loading conditions, a shift from the original stability problem (bifurcation of perfect plates) to a stress problem. Furthermore, plastic yielding in treated plate-structures in the vicinity of laser influenced areas can already occur for rather small external loads. Laser point-pattern treatment on plate-structures introduces only small residual deformations while a mentionable increase in the critical load appears (without exceeding the yield stress during loading within the relevant load-range).

The effect of laser treatment w.r.t. increase of the buckling resistance without mass increase can be evaluated by considering the mass increase by increasing the plate's thickness required for an untreated plate to achieve the same increase of the buckling resistance.

The critical stress for elastic buckling of a plate-structure without residual stresses is given by Eq. 4.5.

$$\sigma^*_{el} = kE\left(\frac{t}{b}\right)^2 \qquad (4.5)$$

In this Eq. 4.5, k is the buckling factor, e.g., taken from [28], for the respective loading and boundary conditions as well as length to width relations $b:a$. E is the Young's modulus, t is the plate-structure's thickness, and b stands for the edge length (with $a = b$ valid for the investigated square plate-structures). For characterising a critical line-load, q^* results from σ^*_{el} by multiplication with the plate-structure's thickness, leading to Eq. 4.6.

$$q^* = kE\left(\frac{t}{b}\right)^2 t \qquad (4.6)$$

Denoting the increase of the critical line-load by laser treatment as $x = \Delta$ and the increase of the thickness, $y = \Delta t$, of the untreated plate-structure for achieving the same critical line load as the treated plate, that means:

$$q^*_1 = q^*_2 \qquad (4.7)$$

than one comes to the following equation:

$$(1+x)kEt^3 = kE[(1+y)t]^3 \qquad (4.8)$$

After rewriting one obtains with Eq. 4.9 a correlation between the increase of the critical load x, of the treated plate, and the necessary increase in an untreated plate-structure's thickness y to result in the same critical load.

$$y = \sqrt[3]{1+x} - 1 \qquad (4.9)$$

For instance, an increase in the critical load, as a consequence of laser-point irradiation, of about 37% equates to a necessary mass increase (due to an increased plate thickness) of about 11% for a residual stress free plate possessing the same buckling resistance. For the case at hand this would be an increase in the plate-structure's thickness from 0.75 mm to 0.83 mm.

Chapter 5

Bead Laying under Optimisation Aspects

After an introduction, this Chapter presents a bead laying algorithm for improving the stability as well as the vibration behaviour of thin-walled structures. Further, the application of this bead laying algorithm on different structures is presented. Finally a conclusion is given. Parts of the work have been published in [10].

5.1 Introduction of Beads into Structures

The introduction of beads represents a further possibility for increasing the buckling resistance as well as the fundamental eigenfrequency of a structure, beside, e.g., thermal treatment as shown in Chapter 4. Beads are specific changes of the structure's geometry in terms of specific locally varying transversal (out-of-plane) deformations. Beads are especially valuable for thin, large-area structures where no global curvature (which automatically implies a stiffening effect) is existent or possible to apply. One of the main advantages of introducing beads instead of applying stiffeners to a structure is the fact, that no additional mass is added. An application example coming from automobile history is shown in Fig. 5.1, whereby beaded plate-metal parts were extensively used.

The effect that contributes to a buckling resistance increase in a beaded structure is not primarily caused by residual stresses but by the geometry of the introduced beads. This geometric effect, due to local and systematic deformation of the plate's original geometry, results in a local rise of the bending stiffness. This stiffening effect is clearly dependent on the direction of the applied load.

Chapter 5 Bead Laying under Optimisation Aspects

Figure 5.1: Commercial vehicle Citroën Typ H (from a time period between 1947 - 1981, depicted model represents a later version) featuring a bodywork with numerous introduced beads. (courtesy of T. Hatzenbichler, LUT, Montanuniversitaet Leoben)

In the literature some contributions can be found, e.g., [18] or [38], which deal with optimised bead introduction according to an increase of the endurable load intensity or bending stiffness as well as an increase of the eigenfrequency of the treated structure. However, papers dealing with bead patterns for increase of buckling resistance are rather rare.

5.1.1 Manufacturing Principles

Two basic possibilities how to introduce beads into a plate-structure are given, see Fig. 5.2. Deep drawing represents a classical and widely spread method for introducing beads into plate-metal parts. Although this method is quite expensive, due to the necessity of complex parts (deep drawing mould, deep drawing punch as well as an optional blank holder), it pays off when it is used for a high number of pieces.

Roll forming of beads, which is also depicted in Fig. 5.2, represents a method which is restricted by the available geometries for the respective beads (in terms of their cross section) but flexible in the possible bead tracks applicable to the plate-metal parts. Practical investigations with the method roll forming were conducted by

Chapter 5 Bead Laying under Optimisation Aspects

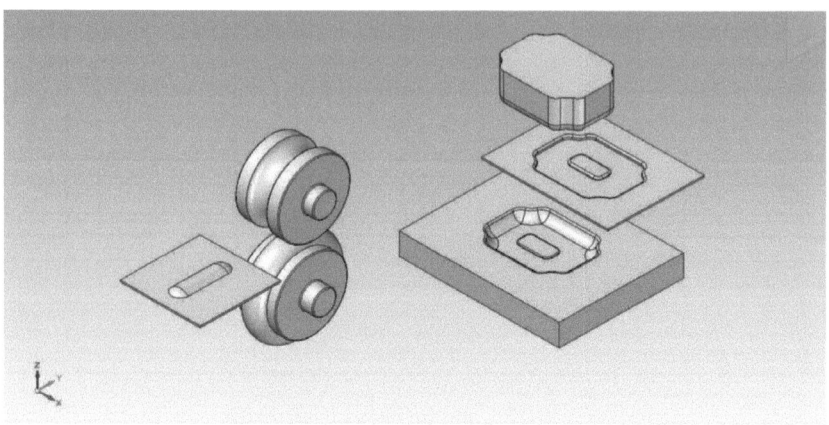

Figure 5.2: Two basic methods for producing beads in plane, half-finished products. Left picture: Roll forming of beads. Right picture: Deep drawing of beads.

the LUT using a machine from the company Trumpf, see [58] for details about the method.

5.2 The Bead Laying Algorithm

The basic idea of the bead laying algorithm comes according to [7] from a concept of lightweight design, namely that stiffeners for increasing the buckling load (or the fundamental frequency) should be positioned so that they disturb the buckling (or vibration) mode most effectively. This means that they should be located at positions where the respective mode of the unstiffened structure exhibits largest curvature. In order to demonstrate this, the example shown in Fig. 5.3 a) should be considered. There, the application of a mid-stiffener-spring changes the buckling mode effectively up to a situation at which the wave number is dublicated and therefore increases the critical load P^*, see load levels in Fig. 5.3 b). The same concept holds for increasing the eigenfrequency when acting on the respective vibration mode.

Chapter 5 Bead Laying under Optimisation Aspects

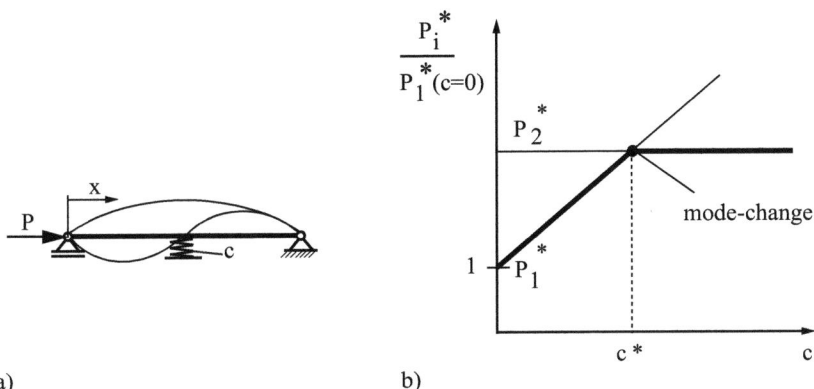

a)   b)

Figure 5.3: Idea of the stiffening concept behind the bead laying algortihm. Picture a): Buckling modes for beam with simple supports at both ends and beam with sufficiently stiff additional mid-stiffener-spring. Picture b): Diagram showing normalised critical load over stiffness of mid-stiffener-spring.

Chapter 5 Bead Laying under Optimisation Aspects

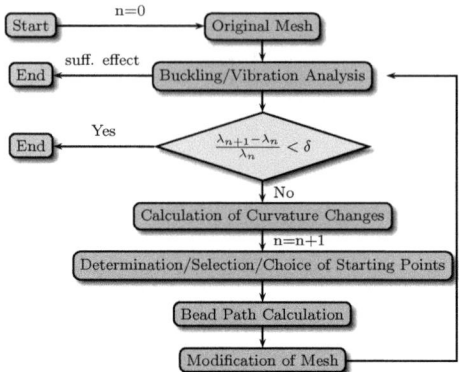

Figure 5.4: Scheme of the incremental bead design procedure.

The manufacturing concept of choice is the above described roll forming, which should enable a flexible introduction of bead pathes into a flat area of a structure. In these plates or shells the position of the maximum absolute values of the principal curvatures of the mode shapes are the relevant positions for effectively stiffen the structure, e.g., by introducing beads. Certainly, introduction of beads change the mode shape and, thus, the distribution of the maximum absolute value of the principal curvatures. Therefore, the bead design according to the described principles requires an incremental procedure. Figure 5.4 shows the concept and the incremental design approach of the bead laying algorithm.

The principal curvature change vector:

The necessary components for determining the principal curvature of buckling modes (or eigenfrequency modes if enhancing the frequency is the goal) are obtained using the FEM programme ABAQUS, see [2]. Discretisation of the investigated structures is conducted by triangular shell elements with bilinear displacement interpolation functions and reduced integration (ABAQUS nomenclature S3R-element, [2]). For all FEM modells, described in the current Chapter, the same material data as already used in Chapter 4 is applied, values at room temperature are taken. In the following the term 'curvature-change' is being used. For already (in the pre-buckling configuration) curved or doubly curved shell structures the term 'curvature of the

Chapter 5 Bead Laying under Optimisation Aspects

buckling mode' is the change of curvature in relation to the unbuckled structure. In the following this term 'curvature change' is, for the sake of simple wording, denoted as 'curvature of the buckling mode' and 'curvature of the vibration mode', respectively.

As result from the FEM programme, curvature components between the undeformed structure and the buckled structure (with a superimposed to the value 1 normalised buckling mode) are delivered. The curvature component tensor for one element is built under usage of the curvature components $SK_x, x = 11, 22, 12$. These components represent the curvature in direction '1', '2' as well as the twist curvature component '12', respectively. The same procedure holds for applying the bead-algorithm to enhance the eigenfrequency behaviour. In that case, the normalised vibration mode is the basis for extracting the curvature components.

The basic procedure of extracting the curvature components and assembling the curvature tensor $\mathbf{A}^{(e)}$, see Eq. 5.1, is shown in Fig. 5.5. In one element (e) all curvature information necessary to assemble the curvature tensor $\mathbf{A}^{(e)}$ is present at the integration point (the used 'S3R' elements have one single integration point).

$$\mathbf{A}^{(e)} = \begin{bmatrix} \kappa_{aa} & \kappa_{ab} \\ \kappa_{ab} & \kappa_{bb} \end{bmatrix} \tag{5.1}$$

With the curvature tensor $\mathbf{A}^{(e)}$, the eigenvalue problem for determining the principal curvatures, as given in Eq. 5.2, follows. Hereby, $\mathbf{E}$ represents the unit matrix.

$$(\mathbf{A}^{(e)} - \kappa_j \mathbf{E}) \vec{\phi}_j = 0 \tag{5.2}$$

Solving Eq. 5.2, one obtains eigenvalues, representing the (absolute) value of the principal curvatures, and respective eigenvectors, representing the direction of the principal curvatures, which are transformed from a local element coordinate system to a global structural coordinate system leading to

$$(\kappa_j, \vec{\phi}_j), j = 1, 2. \tag{5.3}$$

In general, the principal curvature vector is represented by the product

$$\vec{\kappa_j} = \kappa_j \vec{\phi}_j, j = 1, 2 \tag{5.4}$$

Chapter 5 Bead Laying under Optimisation Aspects

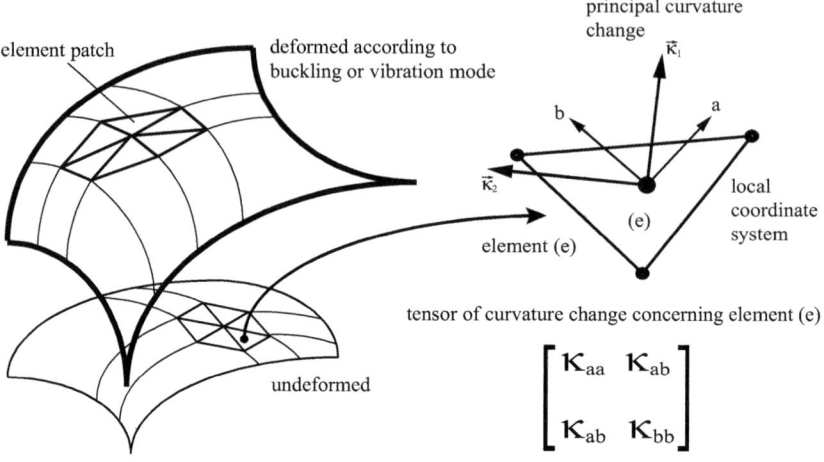

Figure 5.5: Concept of the curvature component extraction to maintain the curvature tensor.

whereby

$$|\vec{\phi}_j| = 1, j = 1, 2 \qquad (5.5)$$

holds.

Investigating one nodal point 'P', the principal curvature vector at this point, $(\widetilde{\vec{\kappa}_j^P})$, is a result of averaging vectors from the surrounding elements, present at the respective integration points (see situation depicted in Fig. 5.6), leading to Eq. 5.6.

$$\widetilde{\vec{\kappa}_j^P} = \frac{1}{\#(e_{patch})} \sum_{(e_{patch})} \vec{\kappa}_j^{(e)}, j = 1, 2 \implies (\widetilde{\kappa_j^P}, \widetilde{\vec{\phi}_j^P}) \qquad (5.6)$$

For the maximal principal curvature one obtains

$$\widetilde{\vec{\kappa}_I^P} = max\left(|\widetilde{\vec{\kappa}_1^P}|, |\widetilde{\vec{\kappa}_2^P}|\right) \vec{e}^{P}_{max}, \vec{e}^{P}_{max} = \widetilde{\vec{\phi}_I^P}. \qquad (5.7)$$

The development of the bead laying path:

Chapter 5 Bead Laying under Optimisation Aspects

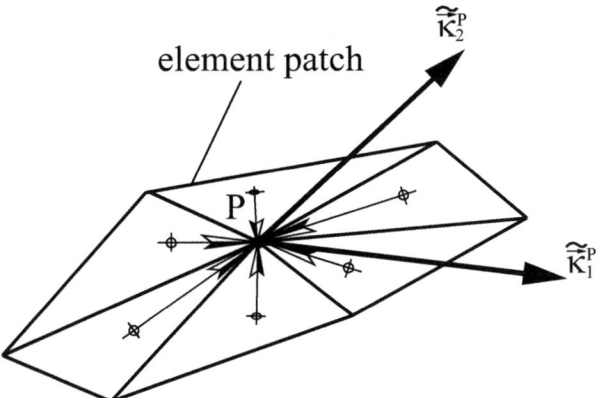

Figure 5.6: Situation on an element patch surrounding a single node. The principal curvature vector is obtained from integration points in each of the surrounding elements and averaged at the node.

Chapter 5 Bead Laying under Optimisation Aspects

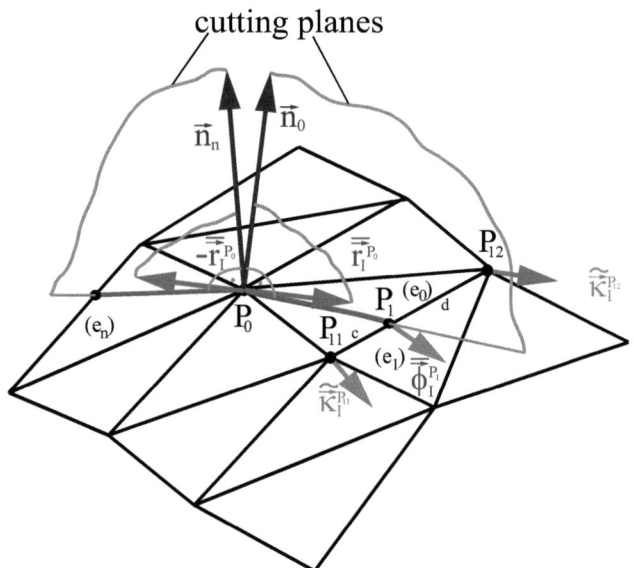

Figure 5.7: Development of a bead laying path.

Chapter 5 Bead Laying under Optimisation Aspects

For the starting point of a bead path (always represented by a nodal point) the vectorial situation, given in Eq. 5.8, is present.

$$\overrightarrow{\widetilde{\kappa_I^{P0}}} = \overrightarrow{\kappa_I^{P0}} = \max_{\forall P} (\overrightarrow{\widetilde{\kappa_I^{P}}}) \qquad (5.8)$$

The nodal point found by the algorithm to act as starting point is used twice because a bead laying direction, notified by the principal curvature vector, is valid in its original direction as well as rotated by 180°, i.e., in the opposite direction. In Fig. 5.7 the situation for the starting point, being a nodal point, is shown with $\pm \overrightarrow{r_I^{P0}}$ pointing into the direction of maximal curvature. Further, planes generated by vectors $\pm \overrightarrow{r_I^{P0}}$ and element normal vectors are applied to cut the affected element edge, leading to the next point (in the current example this is point $P_1^{(e0)}$). For finding a further point in the next element, under usage of the concept shown, vectorial curvature information at adjacent nodes to the current point is used. This is done by interpolation according to Eq. 5.9, leading to curvature data $\overrightarrow{\kappa_I^{P_1}}$. The interpolation scheme is used for all forthcoming points, also if the cutting plane intersection takes place at a nodal point. In that case the new point is artificially shifted to an element's edge to overcome numerical problems.

$$\overrightarrow{\kappa_I^{P_1}} = \frac{\overrightarrow{\kappa_I^{P_{11}}} + (\overrightarrow{\kappa_I^{P_{12}}} - \overrightarrow{\kappa_I^{P_{11}}})c}{c+d} \Longrightarrow (\kappa_I^{P_1}, \overrightarrow{\phi_I^{P_1}}). \qquad (5.9)$$

Again, to obtain the new cutting plane, the vector $\overrightarrow{\phi_I^{P_1}}$, pointing into the direction of maximum principal curvature, is used, see Fig. 5.7. This kind of concept is applied for both bead directions (based on the original starting point - the nodal point). The length of one bead is then the sum of two beads, having started in opposite directions. The maximum length of a bead formed with one increment is given by a predefined value. However additional constraints can limit the length of the bead, too. Length constraints can, for example, become active if bead pathes run into already beaded areas. Predefined 'distance bounds', if defined, can come into action when the bead reaches a 'forbidden' area on the structure. A further length constraint can be the reaching of a structure's edge (e.g., when no 'distance bound' is active).

After fulfilling the amount of layed beads, according to both their amount (predefined number of starting points) and their length (predefined length under consideration of length-constraints), the beaded plate is subjected to the next increment (buckling and eigenfrequency analyses, in current case from $n = 0$ to $n = 1$). As already shown in the algorithm concept, see Fig. 5.4, the number of applied increments of the algorithm depends either on a specific predefined value (buckling load or fundamental frequency) or on a performance definition checking if the improvement in each increment is higher than a predefined value δ, see Eq. 5.10 as well as Fig. 5.4.

$$\frac{\lambda_{n+1} - \lambda_n}{\lambda_n} < \delta \tag{5.10}$$

The complete programme code, containing the bead laying algorithm, which provides the FEM programme with input data for the buckling and vibration analyses, respectively, in each increment is written in Python representing the scripting language of ABAQUS. Medtool[1], see [4], is used as platform programme to opperate the bead laying algorithm. Adaptions to Medtool required for application to the described algorithms according to [4] were used.

5.3 Application of the Bead Laying Algorithm

Two structures are used to test the applicability of the bead laying algorithm. First a flat plate-structure, with the same properties as already used during the laser treatment (Chapter 4), is taken, and further investigations on a U-shaped profile are performed. Details for both structures are given in Tab. 5.1. Boundary conditions applied to each of the two structures are shown in Fig. 5.8.

5.3.1 Application to a Flat Structure - The Plate

In the current case, a one-sided free (three-sided simply supported) plate is subjected to top- and bottom edge-line loading. The resulting normalised buckling mode exhibits maximum values of curvature at the mid of the free edge. The normalised buckling mode and the corresponding distribution of the maximum curvature (depicted as green vectors), present at the untreated plate-structure (increment $n = 0$), as well as the bead resulting from the first increment $n = 1$ (with one chosen starting point) are shown in Fig. 5.9. In the first increment, $n = 1$, the design constraint for

---

[1]Copyright by Assistant Professor Dr. Dieter Pahr, ILSB TU-Vienna 2010.

Chapter 5 Bead Laying under Optimisation Aspects

investigated structure measure	length in mm
square plate-structure	
edge length	280
thickness	0.75
U-shaped profile structure	
base lenght	150
flange height	100
profile length	750

Table 5.1: Dimensions of structures investigated under usage of the bead laying algorithm (plate-structure, U-shaped profile).

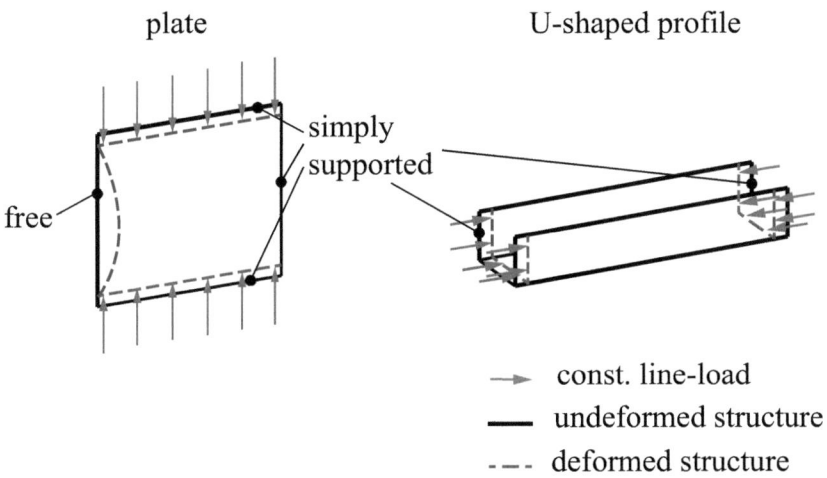

Figure 5.8: Boundary conditions for the two investigated structures (plate-structure, U-shaped profile).

Chapter 5 Bead Laying under Optimisation Aspects

Figure 5.9: First increment of the bead laying algorithm applied to the plate-structure. Left picture: Buckling mode of the untreated plate-structure. Mid picture: Vectors showing maximum curvature values and its respective directions. Right picture: Bead designed in the first increment, shown together with the resulting new buckling mode.

avoiding the choice of a starting point directly situated on the plate's edge (where the highest curvature is located) became active.

For the further development of the beads Fig. 5.10 shows on the left side, for each increment beginning with $n = 2$, the fringe plots of the maximum curvature value (red areas contain high curvature values). On the right side Fig. 5.10 shows the resulting beaded structure. In increment $n = 2$, two starting points were chosen, leading to short beads in the left half of the plate close to the plate's edge and also in the vicinity of the already existing bead. The curvature situation for $n = 2$ is in a way that rather curvature hot spots, not areas of high curvature values, are present. In increment $n = 3$, three starting points are chosen leading to one vertical bead in the mid area of the plate and two inclined beads (mirror image with horizontal axis being the axis of symmetry) in the top and bottom area. In comparison to the situation present during $n = 2$, for $n = 3$ distinct areas of high curvature exist. Last increment $n = 4$ contributes two starting points and therefore two resulting beads which are rather short and located beneath the already existing vertical bead in the plate's mid area.

Because of the reason that the mode structure of the first mode of both buckling and eigenfrequency are nearly identical the obtained bead patterns are suitable both for increasing the critical load concerning buckling and the eigenfrequency. Results achieved by the FEM plate-structures provided with the bead patterns are shown

Chapter 5 Bead Laying under Optimisation Aspects

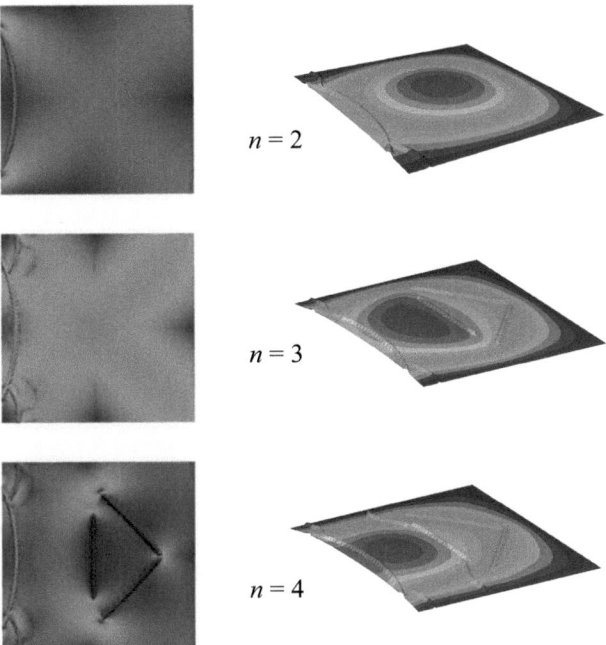

Figure 5.10: Maximum curvature values and resulting beaded plate-structures with the arising first buckling mode shapes being the basis for the next design increment. Depicted are results of the bead laying algorithm for increments $n = 2$ to $n = 4$.

Chapter 5 Bead Laying under Optimisation Aspects

in Fig. 5.11 for the development of the critical load as well as in Fig. 5.12 for the increase in the fundamental frequency for each increment number.

5.3.2 Application to a Profile Structure - The U-shaped Profile

The bead laying algorithm is applied using one time the curvature information from the first buckling mode and one time the curvature information from the fundamental vibration mode. The progress of the design process, fed with first buckling mode based curvature data, is depicted in Fig. 5.13. Hereby, beads are at the present stage of the procedure only present in the profile's flanges. Continuation of the design process shifts areas of largest curvature to the bottom of the U, and, consequently beads appear also there. Furthermore, it is worth mentioning that the half wave number of the buckling mode started with 3. Accordingly three distinct beads per flange were proposed by the algorithm. Due to the stiffening effect by the beads, the half wave number changed to two. Further introduction of beads led to three half waves again.

For the second application of the algorithm, which concerns the fundamental vibration mode, the progress of the incremental procedure is shown in Fig. 5.14. Different to the buckling mode, where especially the profile's flanges (due to their single-sided edge support when comparing to a plate-structure) are showing high curvature values, in the fundamental vibration mode high curvature values are especially present at the bottom of the profile.

Results of both algorithm performances are given in Tab. 5.2.

5.4 Conclusions

The presented bead laying algorithm aims at providing bead patterns for the increase of the buckling resistance as well as the fundamental eigenfrequency of thin-walled structures by introducing incrementally designed bead pathes. As production method roll forming of beads is considered. The basic idea of the algorithm is to apply beads in the vicinity of highest curvature values of the relevant eigenmode (buckling mode, vibration mode) to disturb efficiently the mode shape in order to increase the required potential energy for a given amplitude of the respective mode,

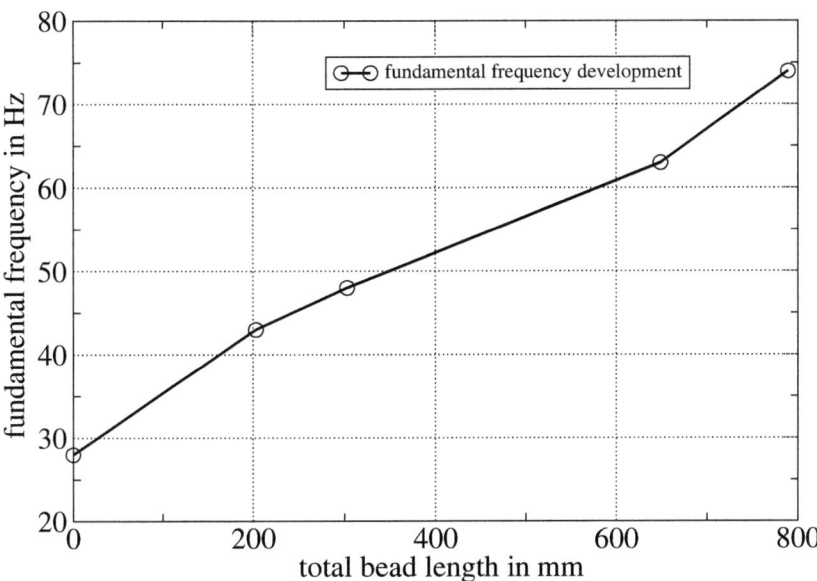

Figure 5.11: Development of the critical load concerning plate buckling after each bead laying increment is applied on the plate-structure.

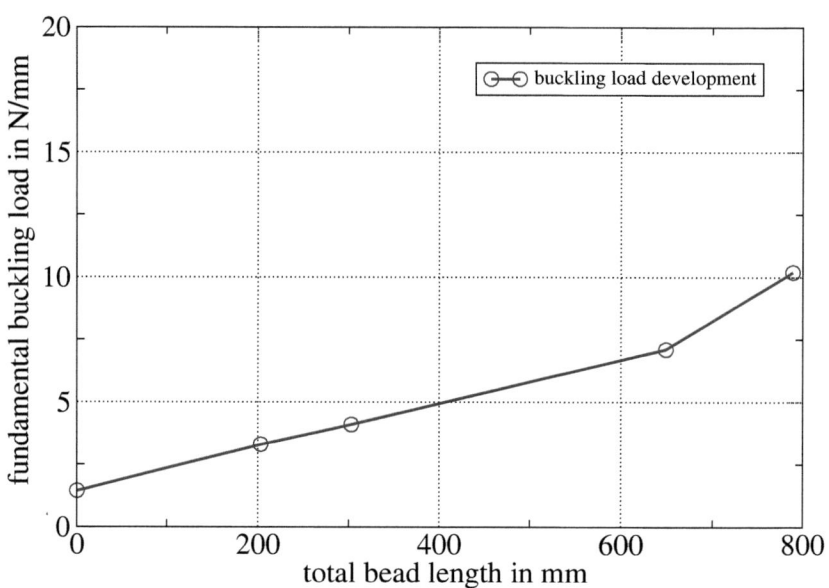

Figure 5.12: Development of the fundamental frequency of the plate after each increment of the bead laying design is applied on the plate-structure.

Chapter 5 Bead Laying under Optimisation Aspects

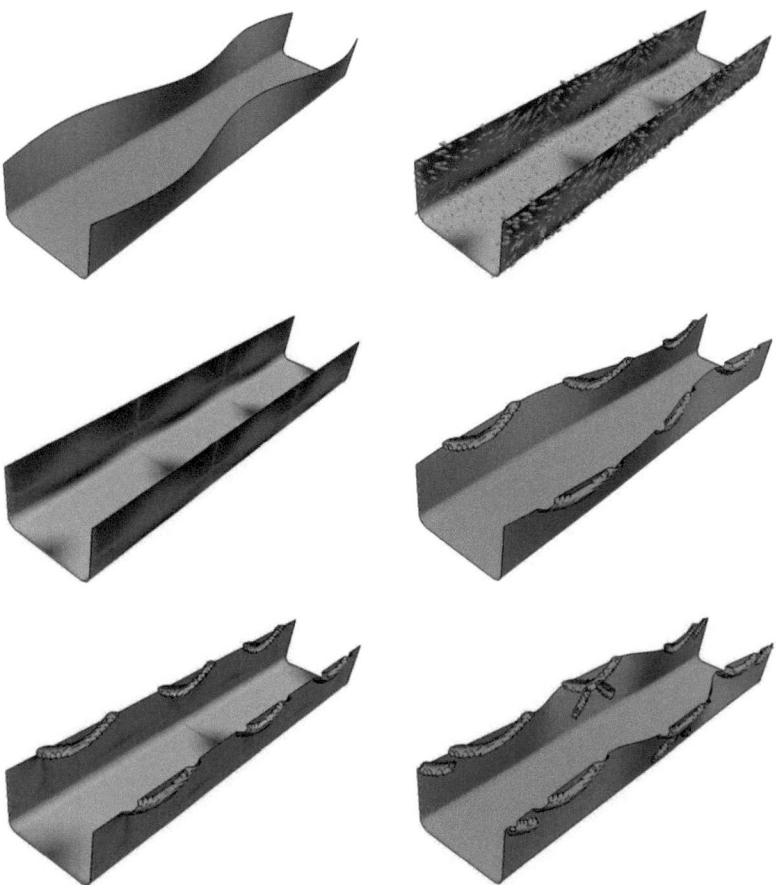

Figure 5.13: Results of the bead laying algorithm applied to the U-shaped profile. Curvature information in each increment is gained from the first buckling mode.

Chapter 5 Bead Laying under Optimisation Aspects

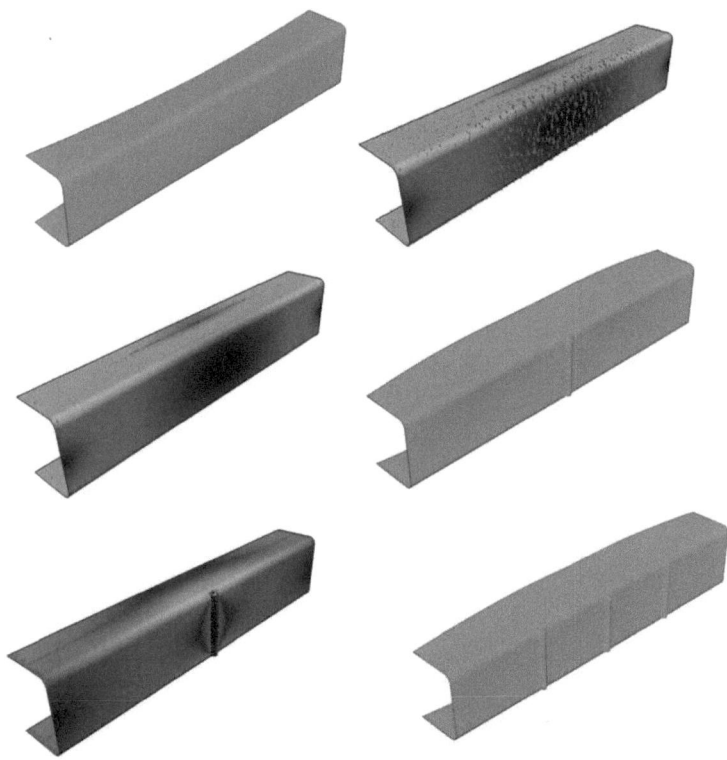

Figure 5.14: Progress of the bead laying designed for the U-shaped profile.

Chapter 5 Bead Laying under Optimisation Aspects

design increment	crit. load in $\frac{N}{mm}$	fund. frequ. in Hz
untreated U-shaped profile	21	49
first buckle mode		
$n = 1$	28	48.8
$n = 2$	36	50
fund. vibration mode		
$n = 1$	22.2	64
$n = 2$	26.6	81

Table 5.2: Results obtained with the bead laying algorithm working on the U-shaped profile under usage of curvature values from the buckling mode and from the vibration mode.

leading to an increased buckling resistance and eigenfrequency, respectively. Main target course in the current case are the first buckling mode as well as the fundamental frequency. If the buckling resistance and/or frequency increase is achieved, the service load can be higher (or a higher safety margin can be obtained if the service load stays the same) as well as resonance through vibrational excitation, occuring during the use of a product, can be easier omitted. Regarding avoidance of higher order resonances the algorithm could also be applied to the corresponding higher order vibration mode.

A significant increase in both buckling load and fundamental frequency is shown for the considered structures after introduction of bead patterns delivered by the algorithm. Especially the investigated plate-structure shows for both targets (buckling load and eigenfrequency) a substantial enhancement with one and the same bead pattern at the same time. Care has to be taken if the shape of the buckling mode and the fundamental vibration mode differ considerable, as it is the case for the U-shaped profile. Therefore, the results obtained for the U-shaped profile are different for buckling load improvement and for improvement of the fundamental frequency. Yet, the presented algorithm is not an otimisation tool but a design tool which accounts for data such as areas of main curvatures - maximum and minimum - and the directions of the main curvatures and proposes bead patterns for enhancing stability and fundamental eigenfrequency in a systematic way.

Chapter 5 Bead Laying under Optimisation Aspects

Weaknesses of the algorithm regard two points. The algorithm needs user intervention (parameter input, such as number of starting points, bead length,...). Furthermore, the measure 'maximum curvature value' may not always be a good choice for a bead starting point position. In some situations the starting point should be chosen within a larger area of higher but not highest curvature values if the highest curvature value corresponds to a 'curvature concentration'. In other words, one should consider an integral of curvatures present in a 'window' of given shape and size. Doing so 'hot spots' of curvatures are not prefered in comparison with areas with somehow smaller curvature distributed over a wider area.

Chapter 6

Roll Forming

This Chapter deals with instabilities which might occur during the production process roll forming and of roll formed products under load. After a detailed literature overview, FEM models are presented to investigate the roll forming process in terms of stability loss. Further FEM models presented herein deal with the calculation of residual stresses in roll formed structures and their effect with respect to buckling under different load cases. Further emphasis is also laid on the numerical investigation of the fundamental eigenfrequency of roll formed structures. A conclusion and summary closes the current Chapter. Work contained in this Chapter is published in [11].

6.1 Introduction

Roll forming is a technical production process which in the majority of cases is performed at room temperature, see [26]. Roll forming is applied for the production of a multitude of profiles in a wide geometric range (e.g., u-profile, pipe-profile or hat-profile). Roll formed parts mostly represent half-finished products and can be found for example in simple products like furnitures, storage racks up to machines in the domestic use. Furthermore roll formed parts are also used in different industrial applications. Feed materials are semi-finished products like metal strips (e.g., taken from coils) or metal plates. In the roll forming process the primary material is being sequently processed in the mill which consists of a number of roll stands. Each roll stand itself consists of two or more counterrotating rolls. During the roll forming process the material thickness has to remain constant on the whole, see, e.g., [14], but areas of reduced material can possibly occur for different profiles. In [31] various mill-lines can be seen. A very detailed description of the roll forming process can be found in [26]. An overview about the suitability of different sheet metal-materials and consequences of different parameters chosen for a roll forming process can be found in [24].

6.2 Literature Review

According to [43] or [56] the motivation for detailed investigations of the roll forming process is the desire to change the parameter finding process for manufacturing suitable products from a 'trial and error' method to a well defined and established procedure. Another reason is to understand the cause of appearing failure in roll formed products. Recent publications also deal with optimisation of existing production lines, e.g., to cut down the number of roll stands, see for example [60]. In the following a number of selected works in the literature with special emphasis on stability loss are described.

6.2.1 Literature - Experimental Data from the Roll Forming Process

Experimental analysis with particular interests in occuring loads and torque during the roll forming process are conducted by [37]. Investigations are made for normal as well as high strength steels for which elastic rebound is a general challenge in forming processes.

6.2.2 Literature - Analytical Methods for Modelling

Early in terms of time and often cited is the work by M. Kiuchi and T. Koudabashi, see for example [33] and [34], using an analytical approach. In [27] a comparison of Kiuchi's approach (M. Kiuchi et al.) with the work of other authors is given. Another analytical approach is presented in [55], investigating different open profiles with regard to resulting longitudinal strains, maximal resulting stresses and resulting geometry of the plate metal's edge.

Chapter 6 Roll Forming

6.2.3 Literature - Modelling based on the Finite Element Method

For modelling of the roll forming process applying FEM, for example in [15], rigid surfaces, representing the rollers, are moved along the strip metal which is at rest are used. A further work, applying combined 2D-3D FEM modelling, is presented in [14] using 2D-modelling with underlying generalised plane strain conditions and 3D-modelling applying shell elements. Between rigid bodies, representing the forming rolls, and the deformable material Coulomb's friction models are used. Describing material behaviour, Hill's model with isotropic hardening is used. Investigated cross-section in this work is trapezoidal and for the 3D FEM model a half-model (with 'longitudinal' symmetry) is applied. State of stress as well as state of strain results from the FEM model during the forming process are compared with experimental results.

According to the author, see [14], longitudinal-strains should stay under a destinct limit and for a reduction of resulting residual stresses in the finished product, forming should preliminary be obtained through bending moments rather than transversal normal forces. This is, according to the author, in contrast to the users' demand of only slight elastic springback to reduce the complexity of calibration.

In [27] roll forming is numerically simulated using a commercial FEM-programme, namely PAM-Stamp. In this case, the solver is of explicit type. A quite detailed literature overview is additionally given by the authors. The process of forming a U-shaped profile is simulated.

A work based on [34], under the use of the Finite Difference Method, is represented by the programme RFPASS, shown in [17]. Results obtained with this programme are compared with experimental data. On one side, fast and numerically efficient solutions can be obtained without the need of applying a commercial FEM programme. But on the other side, related to residual stresses, the calculated values are less than shown by the experiments - according to the authors. Investigations in [17] are based on a U-shaped profile.

Another broad literature overview is given in [43] as well as a simplyfied method, based on the kinematic approach of [34], to simulate profiles with circular cross-section. For these simulations, elasto-plastic material bahaviour with isotropic hard-

Chapter 6 Roll Forming

ening is applied in [43]. Additional comparisons with other works are also done here.

A 2D-simulation, modelling the forming process of pipes with channel-type cross section, is presented in [35].

A simplified 3D-FEM model, under the usage of the programme LUSAS is given in [20]. Discretisation of a half model of the sheet metal is carried out using shell elements. Applying a 'displacement type of loading', as it is called by the authors, the conditions appearing during deformation are simulated in a way that the profile's flanges are increasingly pushed into outwards direction while the profile's web is uniformly pulled. According to the authors, their chosen boundary conditions are in good accordance with experimental observations, and they could exhibit a vanishing longitudinal force. Investigations of U-shaped profiles as well as profiles with circular cross-section are performed as well as a measure for the occurence of flange buckling, due to pressure stresses, called BLS, is introduced by the authors. Comparisons with experimental data as well as with work of other authors is done in [20], too. Possible reasons for occuring failures, regarding the profile, are discussed in this publication. Furthermore the transmittable force from the mill to the sheet metal, which is said by the authors, is another limiting factor for determining the maximal bending angle increase in one mill station.

The commercial FEM software ABAQUS, for explicit analysis, is applied by [56]. In this work especially the occurence of local edge buckling for circular cross section profiles is discussed and thereby initial deformation at the first roll station is seen as particularly critical by the author (degree of deformation at first roll station). The local compressive stresses that occur can be verified by the simulation, as the authors state. These are made responsible for the occurence of the decribed buckling instability. Additionaly, another literature overview is given by these authors. A similar publication, see [57], also shows the importance of a right choice for the degree of the initial deformation.

Another work from 2008, see [32], deals with the deformation process for parts of a slide rail mechanism for which is said that it has to fulfill certain requirements concerning precision. For simulations the FEM software SHAPE-RF is used. Residual strains are investigated by the authors to judge the stability of the product during the roll forming process. Increasing the horizontal distance of two consecutive roll

Chapter 6 Roll Forming

stands as well as a direct reduction of the bending radius for specific critical passes allow a decrease of residual strains and, therefore, the threat of surpassing stability limits. Flower patterns are used to compare two different deformation processes.

Also under the usage of the explicit ABAQUS FEM software, in [59] the simulation of the deformation process of U-shaped profiles is shown. A sensitivity analysis for material parameters is conducted to improve the quality of the resulting end product.

An optimisation of the roll forming process is shown in [60]. A given roll forming mill is optimised according to their angle of elastic rebound as well as their maximal occuring longitudinal strain along the profile's edge as a function of the deformation angle increment and radius of the below situated roll in a roll stand. A further step of optimisation, presented in this work, is to maximise the deformation angle increment in each roll stand to minimise the total number of roll stands. Again ABAQUS in its explicit version is used in this work as FEM software. The optimisation itself is conducted in Matlab.

6.2.4 Literature - Residual Stresses in Formed Products

In [30], residual stresses resulting from roll forming of profiles and their influence on buckling are investigated. These investigations are conducted both analytically as well as experimentally. A more recent publication, see [36], deals with the experimental investigation of residual stresses in hollow profiles. These profiles consist of roll formed profiles which are welded together to obtain hollow sections with a square shaped cross section. Of great interest in these two publications is the exhibiting residual stress, basically the longitudinal stress in direction of the formed profile's axis, along the wall-thickness of the formed product.

6.2.5 Summary of Literature Overview

As shown in recent publications the FEM software ABAQUS in it's explicit version, under usage of shell elements with reduced integration (ABAQUS nomenclature S4R-element), is suitable for detailed simulations of the roll forming process.

Especially residual stresses in the longitudinal direction of the sheet metal should be investigated to judge the issue of buckling stability, both during the forming process and also when the profile is exposed to loads in service. The application of volume elements has not been found in literature yet, but the usage of continuous-shell elements might be possible because of their ability to account for a thickness change (which is not the case for classical shell elements).

6.3 Stability Problems in Connection with the Roll Forming of Profiles

Considering roll formed products, a loss of stability in terms of buckling can already occur in the production process as shown in literature, e.g., [56] or [20]. For wide products with flat profile bases, buckling in the base region is possible (see, e.g., [26]). For narrower profiles a stability loss concerns, in most cases, the profile's flanges. The so called 'flange-buckling' is in that case more likely, see also, e.g., [26].

The kinematics of the forming process 'roll forming' is shown in Fig. 6.1 as an example for a profile exhibiting a half-circular like cross section. In a deformation step points situated on the edge $\overline{AB}$ are subjected to a longer travel than points situated on the, practically non-deformed, profile mid-line $\overline{M_1 M_2}$. At the end of the deformation process only very slight elongation differences should occur in the profile. For the given example, after deformation the length of $\overline{AB}$ should be nearly the same as the length of $\overline{M_1 M_2}$. Due to this, an elongation in the first case has to be followed by a shrinkage at the end of the deformation. As already mentioned, in this case the profile's edges are most affected firstly by tensile stress in the elongation phase followed by compressive stress in the shrinkage phase. These compressive stresses might result in a stability loss for the profile's flanges. Other occuring effects, e.g., in lateral direction, are neglected in the explanation at hand for the appearence of buckling. For roll forming these alternating tensile stress and compressive stress occurence appears between all successive roll stands. Figure 6.2 shows deformations for multiple roll stands where an instability (flange buckling), after the last roll stand, occurs.

Chapter 6 Roll Forming

Additional information, especially for a first estimation of the necessary roll stand number, is for example given in [26]. Strategies to design the 'deformation increment' (i.e., the increment of the bending angle of the flange between two roll stands when forming a U-shaped profile) are also given in [26].

6.3.1 FEM Model representing the Roll Forming Process

In order to study instability phenomena during roll forming, a number of simulations have been performed. They are described in this subsection.

The roll forming mill assembly, used in the explicit ABAQUS[1] simulations, is depicted in Fig. 6.3. Considering the coordinate system, the undeformed sheet metal lies within the y, z-plane and perpendicular to it the x-axis is situated. Generally the model of the mill consists of a preforming- , a forming-, and a postforming-area. The preforming area represents the zone in front of the infeed area into the first mill stand. Here the sheet metal strip is stored between two boundary plates that prevent unwanted movement of the strip while deformation takes place. Within the forming area five horizontal roll stands (direction with respect to the axis of rotation of the used rolls) and one additional vertical roll stand are situated. Parameters of the used rolls are given in Tab. 6.1, additionally general simulation parameters are given in Tab. 6.2.

For a better classification of the deformation process the associated 'roll flower', a pattern showing the stepwise deformation, is depicted in Fig. 6.4. In the model at hand the last deformation step (following the last roll stand) is especially chosen so that the occurence of an instability is provoked.

6.3.2 Results of the FEM Model for the Roll Forming Process Simulation

As already stated before, an instability is provoked by choosing parameters resulting in too high compressive stresses in the profile's edge area. Significant measures for the appearance of buckling instabilities are the existing longitudinal stresses as well as longitudinal strains in the profile's mid-plane along the profile's edges. In

---

[1] ABAQUS/Explicit 6.9, Dassault- Systémes Simulia Corp., Providence, RI, USA.

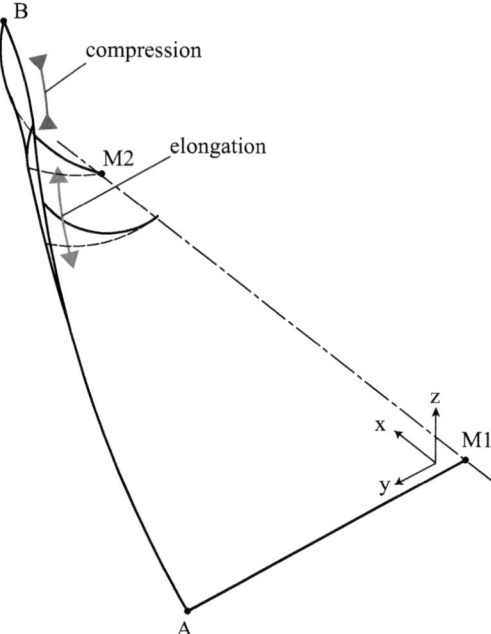

Figure 6.1: Kinematics of the roll forming process, symmetrical half model with semi-circular cross section. Distance $\overline{M_1 M_2}$ keeps constant length during deformation, $\overline{AB}$ edge line during forming step.

Chapter 6 Roll Forming

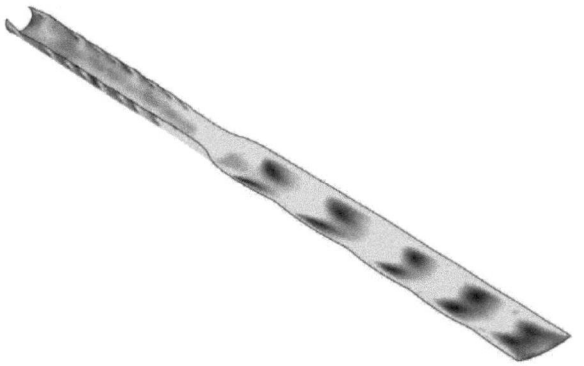

Figure 6.2: Results of a FEM simulation showing step-wise deformation between roll stands. In this model edge buckling occurs after passing through the last roll stand.

Roll stand number	Roll deformation-radius
horizontal 0	120 mm
horizontal 1	100 mm
horizontal 2	80 mm
horizontal 3	60 mm
horizontal 4	50 mm
vertical 0	40 mm

Table 6.1: Dimensions of the used rolls for the roll forming process FEM simulation.

Chapter 6 Roll Forming

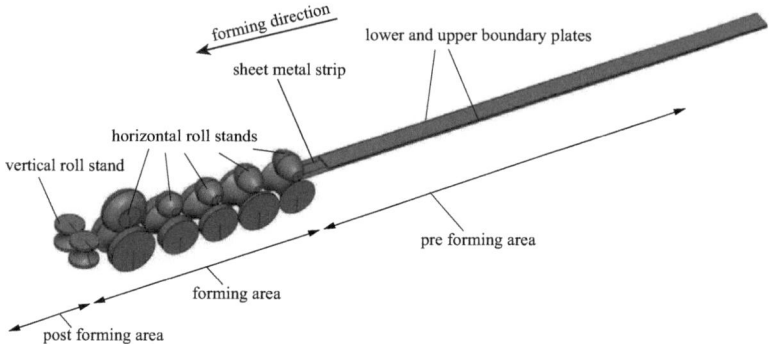

Figure 6.3: Assembly of the FEM model showing the main parts of the modelled roll forming mill to simulate the deformation process (semi-circular cross section of finished product).

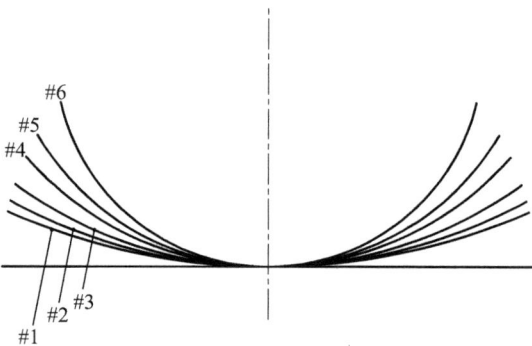

Figure 6.4: Schematic diagram of the 'roll flower' applied in the unstable roll forming process to obtain a semi-circular cross-section profile.

Chapter 6 Roll Forming

Dimensions	
sheet metal length	1500 mm
sheet metal width	100 mm
sheet metal thickness	1 mm
friction coefficient μ	0.1
Mill line data	
velocity of the sheet metal v	1000 mm/s
distance between roll stand horiz. 0 to horiz. 1	180 mm
distance between roll stand horiz. 1 to horiz. 2	180 mm
distance between roll stand horiz. 2 to horiz. 3	180 mm
distance between roll stand horiz. 3 to horiz. 4	210 mm
distance between roll stand vertic. 1 to vertic. 2	215 mm

Table 6.2: Parameters applied in the roll forming process FEM simulation.

Fig. 6.5 the longitudinal stress and in Fig. 6.6 the longitudinal strain, respectively, are depicted (each exhibiting during the last deformation step).

The longitudinal stress σ_{11} within the roll formed profile cannot be seen in terms of a membrane stress. For the completed profile, although undesirably deformed because of occuring instabilities leading to edge-waviness, two element positions are chosen to depict the σ_{11} stress situation along the material's thickness. Both elements are located in the mid span area of the profile. In Fig. 6.7 results for the element at the apex of the cross section are shown. The result shown in Fig. 6.8 comes from an element on the profile's edge.

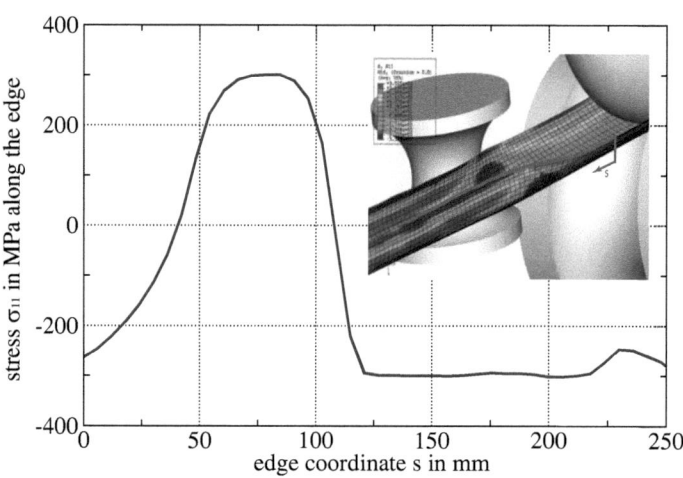

Figure 6.5: Longitudinal stress σ_{11} along profile's edge (path of coordinate s).

Chapter 6 Roll Forming

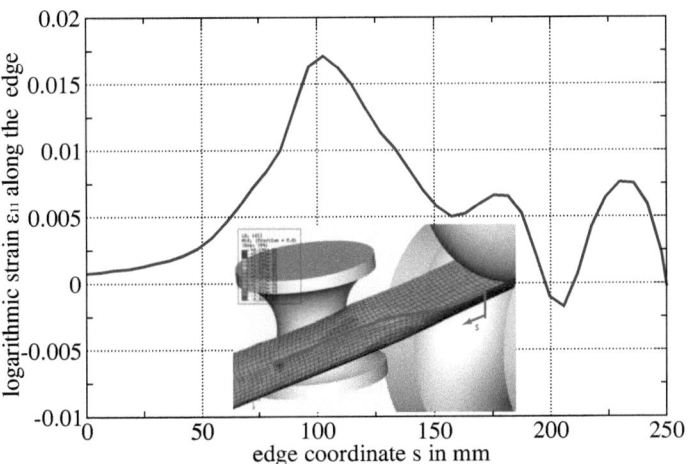

Figure 6.6: Longitudinal strain ε_{11} along profile's edge (path of coordinate s).

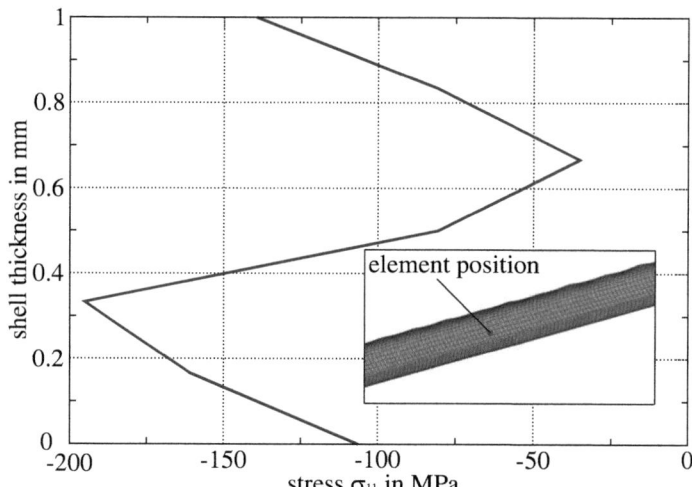

Figure 6.7: Longitudinal stress σ_{11} through the shell thickness for a mid-element.

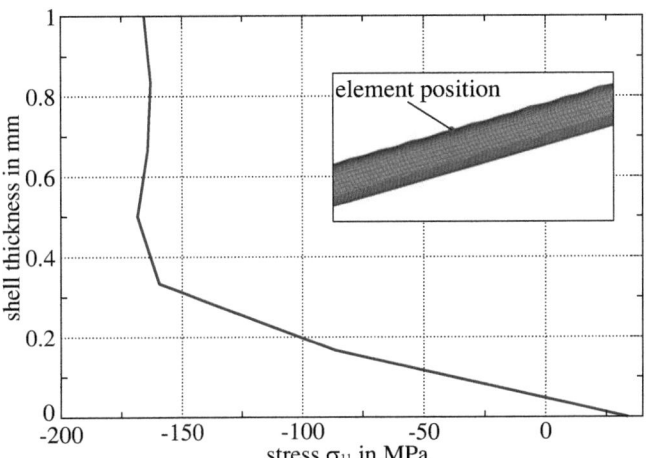

Figure 6.8: Longitudinal stress σ_{11} through the shell thickness for an edge-element.

6.4 Influence of Residual Stresses due to Roll Forming on the Stability of the Rollformed Product

Results obtained in Chapter 6.3 show that care has to be taken for dimensioning a roll forming line for the treatement of thin plate metal material due to the threat of occuring instabilities. Also if no instabilities occur during the roll forming process, depending on the chosen process parameters, significant residual stresses are comprised in the finished product. Load carrying capacities of finished products are strongly dependent on such residual stresses.

In the following for a U-shaped profile it's production (by roll forming) and it's performance under loading are simulated.

6.4.1 Roll Forming Mill for an U-shaped Profile

For simulating the roll forming of a U-shaped profile a rather conservative design, see Fig. 6.9, is chosen to overcome the occurence of instabilities during the roll forming process in any case. Therefore, both the degree of deformation increase between to roll stands (chosen to be low) as well as the distance between two consecutive roll stands (chosen to be rather wide) should assure an avoidance of instabilities.

Details concerning the deformation-increase during the roll forming process are given in Tab. 6.3 and a flower diagram is depicted in Fig. 6.10. Further, values of simulation parameters, such as original sheet metal dimensions, friction coefficient, process velocity and mill stand distances are given in Tab. 6.4.

Figure 6.11 shows the U-shaped profile during the roll forming simulation. Although the parameters for the roll forming process simulation were chosen with care the obtained U-profile was not perfect, as depicted in Fig. 6.12. Especially within the intake-zone of the profile as well as in the zone mainly affected during pullout of the profile out of the forming mill uneven deformations are noticeable. For investigations of a roll formed profile, containing a more or less even geometry and also an even stress-distribution, front and end pieces are cut away.

Chapter 6 Roll Forming

Figure 6.9: Assembly of the FEM model showing the roll forming mill to simulate the production of the U-shaped profile.

Roll stand number	U-profile flange angle
R 0	10°
R 1	25°
R 2	40°
R 3	55°
R 4	70°
RR 1	85°
RR 2	90°
RR 3	95°

Table 6.3: Dimensions of the flange angle for the FEM model to simulate the production of the U-shaped profile.

Chapter 6 Roll Forming

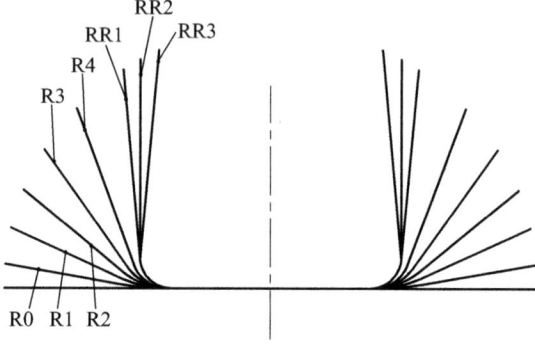

Figure 6.10: Schematic diagram of the 'roll flower' applied in the simulation of the production of the U-shaped profile.

Dimensions	
sheet metal length	900 mm
sheet metal width	135 mm
sheet metal thickness	1 mm
friction coefficient μ	0.1
Mill line data	
velocity of the sheet metal v	1000 mm/s
distance between roll stand R0 to horiz. R1	500 mm
distance between roll stand R1 to horiz. R2	500 mm
distance between roll stand R2 to horiz. R3	500 mm
distance between roll stand R3 to horiz. R4	500 mm
distance between roll stand R4 to horiz. RR1	500 mm
distance between roll stand RR1 to vertic. RR2	500 mm
distance between roll stand RR2 to vertic. RR3	500 mm

Table 6.4: Parameters applied in the FEM simulation of the roll forming process for the production of the U-shaped profile.

Figure 6.11: U-shaped profile during the roll forming process simulation.

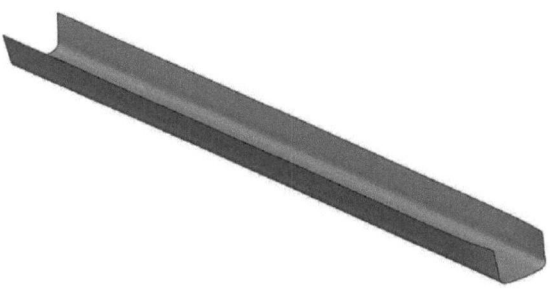

Figure 6.12: Resulting U-shaped profile from the roll forming process simulation.

Chapter 6 Roll Forming

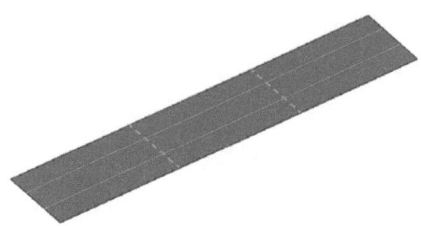

Figure 6.13: Partitioned plate metal structure representing the basic product for the roll forming of the U-shaped Profile. Longitudinal dashed lines show bending zone of profile's flanges, horizontal dashed lines show cutting pathes.

6.4.2 Cutting of the U-shaped Profile

After the completed deformation process, investigations concerning energy measures are performed, to assure that a state of equilibrium has set in. Afterwards, the profile is cut at both ends to become one-third of its length (the middle part of the profile is taken for further investigations). In terms of simulation technique the elements in the FEM model which are cut are simply deactivated. The original plate metal structure, with respective partitions, is depicted in Fig. 6.13. The simulation of the cutting process (following the simulation of the forming process) is performed in a nonlinear analysis in order to obtain the stress redistribution. Subsequently linear buckling and eigenfrequency analyses are performed. The distributions of the longitudinal stresses σ_{11}, before the cutting operation and afterwards, are depicted as fringe plots in Fig. 6.14.

After the cutting is carried out and the stress redistribution is set, stress distributions of longitudinal stresses σ_{11} in the respective surface (plate's top surface, plate's mid-

Chapter 6 Roll Forming

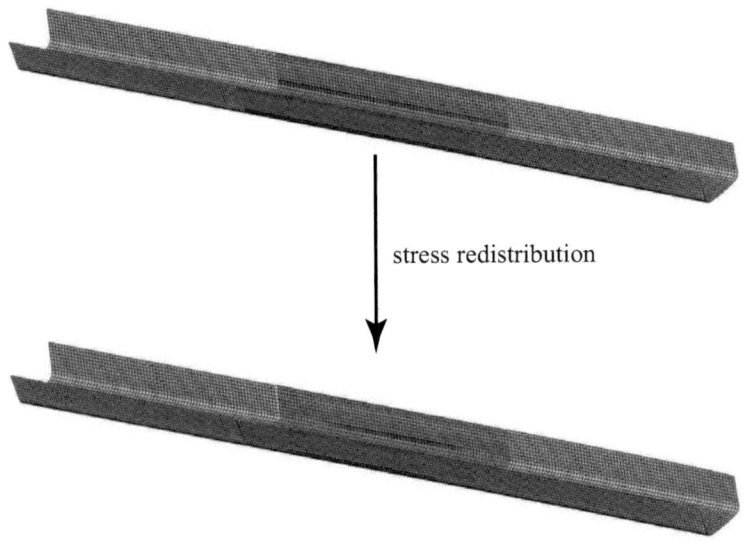

Figure 6.14: Fringe Plot of longitudinal residual stresses σ_{11} before the cutting operation and afterwards, respectively (grey elements are deactivated).

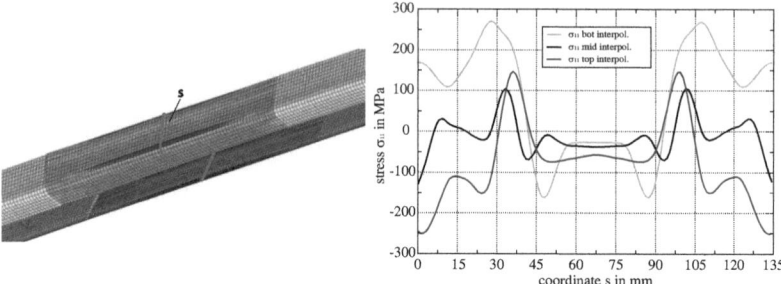

Figure 6.15: Distribution of longitudinal residual stress σ_{11} along the path s following the cross section in the mid span. Stresses are shown for profile's top-surface, mid-surface and bottom-surface.

plane and plate's bottom surface), for a cutting line perpendicular to the profile's axis (runnning through the profile's midpoint), are shown in Fig. 6.15.

6.4.3 Comparison with 'perfect' Residual Stress free Profile

Two different types of performance comparisons are investigated for the U-shaped profile. On one side buckling analyses under longitudinal compressive load and eigenfrequency analyses are conducted, and on the other side bending loads are applied to investigate the collapse behaviour under transverse load. In both cases, a comparison between the roll formed profile (with residual stresses) and a 'perfect' profile (without residual stresses) is conducted.

Buckling Behaviour

For determining axial compressive loads, which are critical in terms of buckling, linear eigenvalue analyses, with boundary conditions as shown in Fig. 6.16, are performed.

Chapter 6 Roll Forming

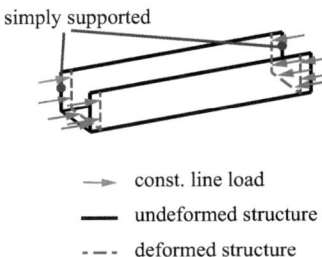

Figure 6.16: Boundary conditions applied during the linear buckling analysis.

A direct comparison between the first buckling mode of the roll formed U-shaped profile and the 'perfect' U-shaped profile shows hardly any difference, see Fig. 6.17. Regarding the critical load, the residual stress affected profile shows a decrease by 16 % compared to the 'perfect' profile.

Eigenfrequency Behaviour

As an important and representating dynamic parameter of the structure, also in this case, the fundamental frequency is investigated. Analysis are done under usage of boundary conditions depicted in Fig. 6.18. It holds also for the fundamental frequency (as it is the case for the first buckling mode) that no significant difference can be observed, see Fig. 6.19. As for the critical load, concerning buckling, also for the fundamental frequency a decrease is noticeable, for the current case by 6 %.

Collapse Behaviour under Bending

For both perfect and roll-formed U-shaped profiles the collapse behaviour under bending is investigated. This kind of static loading type is performed by applying a transversal displacement to the free end of the one-sided clamped U-shaped profile. The prescribed displacement is set to be in the direction of the axis of symmetry of the cross section. In one case (load case 1) the prescribed displacement points upwards in the other case (load case 2) it points downwards. In Fig. 6.20 the boundary

Chapter 6 Roll Forming

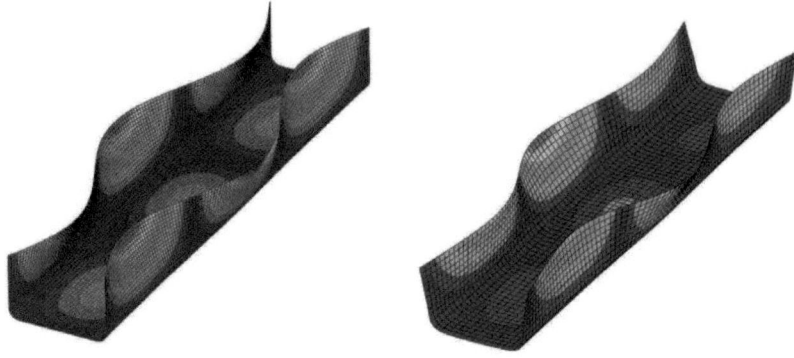

Figure 6.17: First buckling mode. Left picture: 'perfect' U-shaped profile. Right picture: U-shaped profile containing residual stresses due to roll forming.

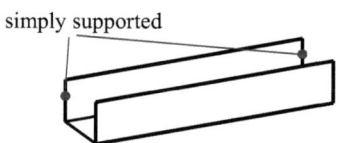

Figure 6.18: Boundary conditions applied during the linear eigenfrequency analysis.

Chapter 6 Roll Forming

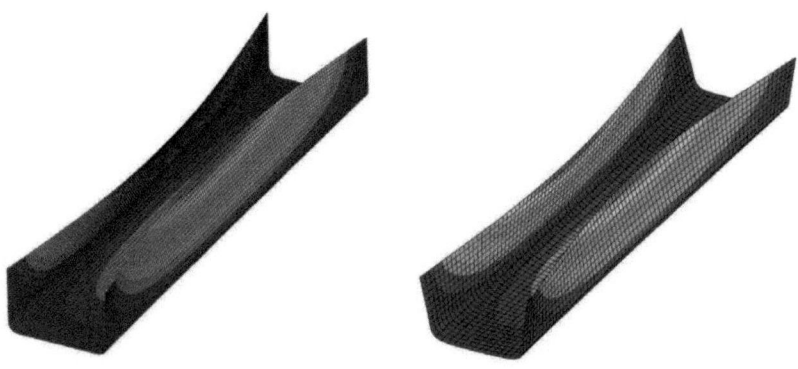

Figure 6.19: Fundamental vibration mode. Left picture: 'perfect' U-shaped profile. Right picture: U-shaped profile containing residual stresses due to roll forming.

conditions, with respect to support and prescribed displacement (representing the load), are shown in detail.

Load case 1

For a prescribed displacement of 30 mm (load case 1: displacement in upwards direction), resulting in an upwards bending of the U-shaped profiles, the behaviour of the 'perfect' profile and that of the roll formed profile are compared according to their deformation and finally to their collapse behaviour, see Fig. 6.21. For the perfect U-shaped profile flange buckling sets in during deformation and finally results in a postbuckling deformation state present in the picture shown in Fig. 6.21. Contrary to the obvious deformations appearing at the flange, the perfect U-shaped profile's base exhibits only slight deformations in the first third of the overall profile length (measured from the profile's end position where clamped boundary conditions are acting).

A different behaviour is shown in case of the roll formed profile subjected to load case 1 (a prescribed upwards displacement). Deformations occuring, while the pre-

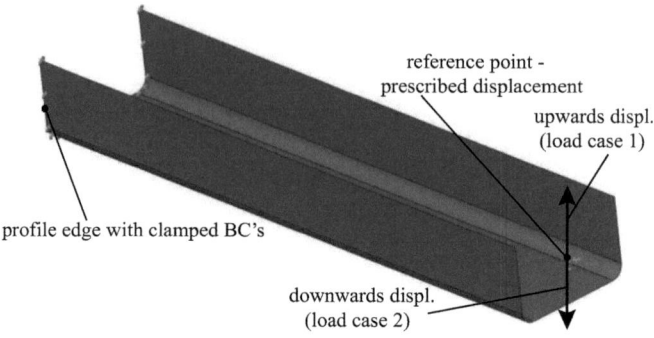

Figure 6.20: Boundary conditions for the collapse behaviour investigation for the U-shaped profile.

scribed displacement is acting, are at first similar to the 'perfect' profile subjected to load case 1 meaning that two areas of local deformations appear. During an increase of the prescribed displacement the resulting profile-deformations, which are present at the roll formed profile's flanges close to the clamped area (facing inwards with respect to the profile), decrease again. The roll formed profile exhibits less stiffness compared to the 'perfect' profile within this load case 1 because of compressive residual stresses included in the profile's flange areas to which the compressive stresses due to displacement induced bending are additionally acting. Therefore, plastic deformations occur as limiting factor resulting in a reduced stiffness (slope of the load displacement curve within the stable domain). The post-buckling deformations for the roll formed profile are more localised. The resulting load displacement curves are shown for both profiles in Fig. 6.23.

Load case 2

In load case 2 a displacement of 30 mm in downwards direction is prescribed. Figure 6.22 shows both 'perfect'- and roll formed U-shaped profiles subjected to load case 2 at the end of the prescribed displacement. During load increase the perfect U-shaped profile reaches considerable local plastic deformations within the profile's flanges (appearing in the vicinity of the profile's clamped end). These plastic deformations appear before the stability limit is reached. Due to this, already in the

Chapter 6 Roll Forming

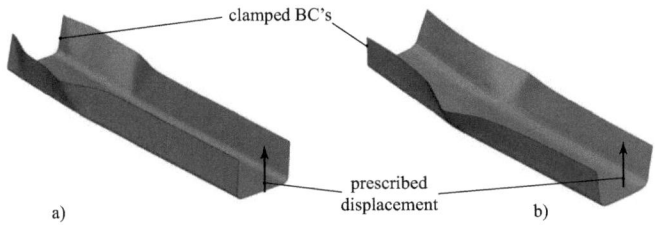

Figure 6.21: Collapse behaviour of U-shaped profiles subjected to load case 1 (a prescribed upwards displacement). Picture a): 'perfect' U-shaped profile. Picture b): roll formed profile containing residual stresses.

pre-buckling domain, a clear nonlinear behaviour is visible in the load displacement behaviour, as Fig. 6.23 shows. In the post-buckling domain a deformation with increasing extent and finally leading to the cause of the collapse is situated in the profile's base (also in the vicinity of the profile's clamped end). On the contrary, the profile's flanges, which are subjected to tensile stresses, remain nearly undeformed. The collapse appears at a prescribed displacement of 4.5 mm. Compared to the same perfect U-shaped profile, subjected to load case 1, a clear increase in the collapse load is present. This increase results from the respective stress situation in the concerned parts of the profile. In load case 1 buckling relevant compressive stresses, resulting from the load, are predominantly present in the profile's flanges. In load case 2 they are present in the profile's base. Comparing both profile's flanges and profile's base to flat plate-structures under compressive load, in terms of buckling the base is advantageous compared to the flange beacause of it's two-sided support compared to the on-sided support existing for the flanges. This simplified comparison explains the reason for the increased collapse load in present case.

For the U-shaped profile with residual stresses due to roll forming a quite similar behaviour as for the 'perfect' profile is present when subjected to load case 2. After exceeding the collapse load, additional but small deformations in the area close to the clamped side occur and main deformations in the base are farer away from the clamped side compared to the 'perfect' profile. The main difference to the 'perfect' profile is a far higher collapse load level for the roll formed profile. On

Chapter 6 Roll Forming

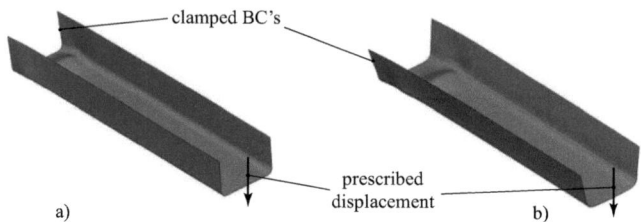

Figure 6.22: Collapse behaviour of U-shaped profiles subjected to load case 2 (a prescribed downwards displacement). Picture a): 'perfect' U-shaped profile. Picture b): roll formed profile containing residual stresses.

average, compressive residual stresses are present in the roll formed profile's flanges. Subjected to load case 2 dominant tensile stresses counteract these predominant compressive stresses in the flanges. Thus, the residual stresses present in the profile give an explanation for the rise in the collapse load for this kind of loadcase.

6.5 Summary and Conclusion

Roll forming represents a widely spread production process for manufacturing long profiles. Especially for profile-raw material with a small thickness, stability problems during the roll forming process may arise. One example of such a stability loss is the occurence of edge buckling. Relevant parameters which directly influence these occurence are the distance between two roll mill stands and the deformation increase per mill stand. By all means a stability loss during the production process has to be avoided because otherwise the resulting product cannot be used in most cases.

The finished roll formed product contains, as simulations show, both a specific residual stress distribution over its width and its thickness. Additionally, both profile's ends are influenced by the feed in procedure as well as the ejection procedure of the profile during the roll forming resulting in a noneven residual stress distribution in the vicinity of the ends. Cutting out of pieces may be a possible solution to enhance

Chapter 6 Roll Forming

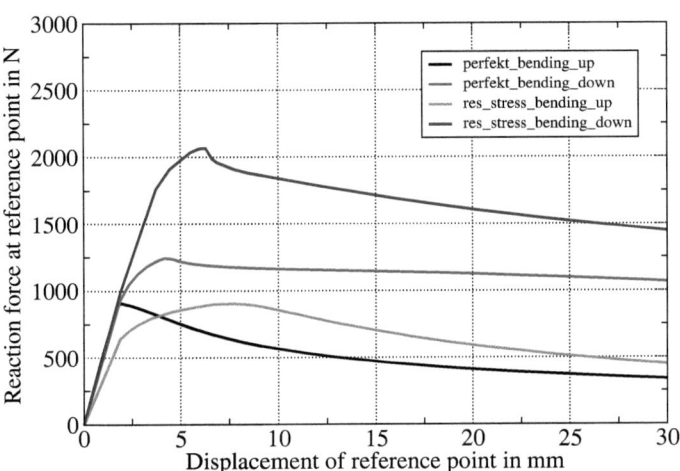

Figure 6.23: Load displacement curves for both load cases, each for the 'perfect' U-shaped profile as well as for the roll formed imperfect and residual stress containing profile.

Chapter 6 Roll Forming

the evenness of the longitudinal stress distribution at the profile's ends. This has been done in the simulations, too. For the cutted roll formed profile containing residual stresses and deformations (in a way that the desired U-shape of the finished profile is not exactly reached) both buckling load and eigenfrequency are decreasing in a direct comparison to a perfect, stress-free U-shaped profile. Positive effects from the residual stresses are expected if they are present predominantly as membrane stresses.

A different result is obtained from the investigated collapse behaviour under bending loading. Depending on the direction of bending the residual stresses may lead to an increase or a decrease in the collapse load.

List of Figures

2.1 Comparison of load-displacement behaviour of different (perfect an imperfect) plate-strutures. 13

2.2 Example for the dependency of the effective length l_K on the boundary conditions of a bar. Case a) one side free, case b) one side guided support. 14

2.3 Schematic example of a Southwell diagram. Input: a displacement under load, P load applied on structure. Output: P_K^* structure's critical load, a_0 predeformation of the structure. 16

3.1 Boundary and loading conditions present at the T-joint FEM model. 20

3.2 Deformation procedure on an T-joint connection. Picture a): Unloaded state. Picture b): Stress state after full prescribed displacement, point C still held in position. Picture c): Residual stress state after complete relief of the prescribed displacement. (Valid for all three pictures: von Mises stresses in MPa, each in the structure's mid-plane) 21

3.3 Residual stress state (von Mises stresses in MPa) after embossing at three positions in a plate-structure. Stress state depicted in the structure's mid-plane. 22

3.4 Top picture: Temperature field (in °C at the plate-structure's top surface) during continuous laser irradiation along a straight laser track. Bottom picture: Corresponding von Mises stress field (in MPa) occuring at the same time. 23

3.5 High speed milling machine for applying the borehole method to measure residual stresses. (courtesy of G. Figala, LUT, Montanuniversitaet Leoben) 26

List of Figures

3.6 X-ray diffractometry for measuring residual stresses in a non-destructive way at the MCL (Materials Center Leoben, Forschung GmbH). (courtesy of G. Figala, LUT, Montanuniversitaet Leoben) 26

4.1 Material properties of mild steel type DC01 over a temperature range up to 800°C, taken from [51]. 30

4.2 Material flow curves of mild steel type DC01 at specific temperatures due to hot tensile tests performed at the LUT in Leoben. (courtesy of G. Figala, LUT, Montanuniversitaet Leoben) 31

4.3 Boundary conditions of the square plates (treated with a single laser dot) with $a_0 = 90$ mm used in the FEM analyses to investigate the residual stress development during heat treatment. 34

4.4 Fringe-plot showing the distribution of σ_{yy}-stresses in MPa of plate type W2B1.1, treated 0.5 s with a single laser point. Depicted stress state is present in the plate's mid-plane after complete cooling down. 35

4.5 Horizontal distribution of σ_{xx}-stresses in MPa, plate's top-surface position, while laser irradiation is activated (heat input), plate type W2B1.1. 36

4.6 Horizontal distribution of σ_{xx}-stresses in MPa, plate's top-surface position, during cooling down, plate type W2B1.1. 37

4.7 Horizontal distribution of σ_{yy}-stresses in MPa, plate's top-surface position, while laser irradiation is activated (heat input), plate type W2B1.1. 38

4.8 Horizontal distribution of σ_{yy}-stresses in MPa, plate's top-surface position, during cooling down, plate type W2B1.1. 39

4.9 σ_{yy}-stress distribution in MPa, along a horizontal path, on the plate's top-surface. The horizontal path advances through the plate midpoint, plate type W2B1.1. 40

4.10 σ_{yy}-stress distribution in MPa, along a horizontal path, in the plate's mid-plane. The horizontal path advances through the plate midpoint, plate type W2B1.1. 41

4.11 σ_{yy}-stress distribution in MPa, along a horizontal path, on the plate's bottom surface. The horizontal path advances through the plate midpoint, plate type W2B1.1. 42

List of Figures

4.12 σ_{yy}-stress distribution in MPa, along a vertical path, on the plate's top-surface. The vertical path advances through the plate's midpoint, plate type W2B1.1. 43

4.13 σ_{yy}-stress distribution in MPa, along a vertical path, in the plate's mid-plane. The vertical path advances through the plate's midpoint, plate type W2B1.1. 44

4.14 σ_{yy}-stress distribution in MPa, along a vertical path, on the plate's bottom surface. The vertical path advances through the plate's midpoint, plate type W2B1.1. 45

4.15 Details of σ_{yy}-stress distribution, along a horizontal path on the plate's top-surface. The horizontal path advances through reference point no. 2, plate type W2B1.1. 46

4.16 Details of σ_{yy}-stress distribution, in the plate's mid-plane. The horizontal path advances through reference point no. 2, plate type W2B1.1. 47

4.17 Details of σ_{yy}-stress distribution, along a horizontal path on the plate's bottom surface. The horizontal path advances through reference point no. 2, plate type W2B1.1. 48

4.18 Distribution of σ_{yy}-stresses in MPa along the plate's thickness coordinate in reference point no. 1 (plate's midpoint), plate type W2B1.1. 50

4.19 Distribution of σ_{yy}-stresses in MPa along the plate's thickness coordinate in reference point no. 2, plate type W2B1.1. 51

4.20 σ_{yy}-stress distribution in MPa, horizontal path, on the plate's top-surface, plate type W2B1.2. 52

4.21 σ_{yy}-stress distribution in MPa, horizontal path, in the plate's mid-plane, plate type W2B1.2. 53

4.22 σ_{yy}-stress distribution in MPa, horizontal path, on the plate's bottom surface, plate type W2B1.2. 54

4.23 σ_{yy}-stress distribution in MPa, vertical path, on the plate's top-surface, plate type W2B1.2. 55

4.24 σ_{yy}-stress distribution in MPa, vertical path, in the plate's mid-plane, plate type W2B1.2. 56

4.25 σ_{yy}-stress distribution in MPa, vertical path, on the plate's bottom surface, plate type W2B1.2. 57

4.26 σ_{yy}-stress distribution in MPa, along the plate's thickness in the plate's midpoint, plate type W2B1.2. 58

List of Figures

4.27 σ_{yy}-stress distribution in MPa, along a horizontal path, on the plate's top-surface, plate type W2B4.1. 60

4.28 σ_{yy}-stress distribution in MPa, along a horizontal path, in the plate's mid-plane, plate type W2B4.1. 61

4.29 σ_{yy}-stress distribution in MPa, along a horizontal path, on the plate's bottom surface, plate type W2B4.1. 62

4.30 σ_{yy}-stress distribution in MPa, along the plate's thickness in the plate's midpoint, plate type W2B4.1. 63

4.31 Displacement normal to plate-plane U3 (direction z-axis) in mm, existing in the plate after heat input and complete cool-down. (Deformation scale factor: 100), plate type W2B1.2. 65

4.32 Simulation showing a 'foldover'-like behaviour with deformation direction changing during laser irradiation. For enhancement of visibility reason fringe plots of von Mises equivalent stresses in MPa (which exhibit high values within the laser influencing area) are shown (situation on the plate's bottom surface). (Deformation scale factor: 20). 66

4.33 Sketch of the general loading situation of the laser treated square plate-structures (dimensions: $a_1 = 280$ mm, thickness 0.75 mm). Denoted directions x, y or z, with the plate lying in the x,y-plane, show a blockage in respective translatoric direction. 67

4.34 Sketch of introduced laser track geometries and point pattern. Continuous circle-, I- and X- geometry as well as point pattern for the pointwise laser treatment. 68

4.35 Boundary conditions of the square plates with $a_1 = 280$ mm and a thickness of 0.75 mm. a) Boundary conditions used during the FEM analyses for the laser heat input and cooling down. b) Boundary and loading conditions being active during buckling- and eigenfrequency analyses. Denoted directions x, y or z, with the plate lying in the x,y-plane, show a blockage in respective translatoric direction. 69

4.36 Fringe plot of temperature field in °C at the plate's top-surface, heat input due to moved laser source, circle geometry. 70

4.37 Fringe Plot of residual stresses σ_{yy} in MPa situated in the plate's mid-plane, heat input due to moved laser source, circle geometry. .. 71

List of Figures

4.38 σ_{yy}-stress distribution in MPa, along $y \equiv 0$, after heat treatment with moving laser source, circle geometry. Stresses are situated in the plate's mid-plane. 72

4.39 Fringe Plot of residual out-of-plane displacements U3 in mm. Heat treatment with moving laser source, circle geometry. 73

4.40 Residual out-of-plane displacement in mm along $y \equiv 0$. Heat treatment with moving laser source, circle geometry. 74

4.41 Fringe Plot of residual out-of-plane displacements U3 in mm. Isometric view, deformation scale 10. Heat treatment with moving laser source, circle geometry. 75

4.42 Fringe plot of temperature field in °C situated at the plate's top-surface, heat input due to moved laser source, X-pattern geometry. . 76

4.43 Fringe Plot of residual stresses σ_{yy} in MPa situated in the plate's mid-plane, heat input due to moved laser source, X-pattern geometry. 77

4.44 σ_{yy}-stress distributions (in MPa), along $y \equiv 0$, after heat treatment with moving laser source (X-pattern geometry). Stresses are situated in in the plate's mid-plane. 78

4.45 Fringe Plot of residual out-of-plane displacements U3 in mm. Heat treatment with moving laser source, X-pattern geometry. 79

4.46 Residual out-of-plane displacements U3 in mm, along $y \equiv 0$. Heat treatment with moving laser source, X-pattern geometry. 80

4.47 Fringe Plot of residual displacements U3 in mm under isometric view, deformation scale 10. Heat treatment with moving laser source, X-pattern geometry. 81

4.48 Fringe plot of temperature field in °C at the plate's top-surface, heat input due to moved laser source, I-pattern geometry. 82

4.49 Fringe Plot of residual stresses σ_{yy} in MPa in the plate's mid-plane, heat input due to moved laser source, I-pattern geometry. 83

4.50 σ_{yy}-stress distribution (in MPa) in the plate's mid-plane along $y \equiv 0$ after heat treatment with moving laser source, I-pattern geometry. . . 84

4.51 Fringe Plot of residual displacements U3 in mm. Heat treatment with moving laser source, I-pattern geometry. 85

4.52 Residual out-of-plane displacements U3 in mm along $y \equiv 0$. Heat treatment with moving laser source, I-pattern geometry. 86

List of Figures

4.53 Fringe Plot of residual out-of-plane displacements U3 in mm, deformation scale 10. Heat treatment with moving laser source, I-pattern geometry. 87

4.54 Sequence of point position's for laser treated plate type 25P. 87

4.55 Fringe plot of temperature field in °C at the plate's top-surface, heat input due to pointwise laser treatment, 25P-pattern geometry. 88

4.56 Fringe Plot of residual stresses σ_{yy} in MPa situated in the plate's mid-plane, heat input due to pointwise laser treatment, 25P-pattern geometry. 89

4.57 Residual σ_{yy}-stress distribution in MPa along $y \equiv 0$ in the plate's mid-plane after discontinuous, pointwise laser treatment (25P-pattern geometry). 90

4.58 Fringe Plot of residual out-of-plane displacements U3 in mm. Heat input due to pointwise laser treatment, 25P-pattern geometry. 91

4.59 Residual out-of-plane displacements U3 in mm along $y \equiv 0$. Heat input due to pointwise laser treatment, 25P-pattern geometry. 92

4.60 Fringe plot of residual out-of-plane displacements U3 in mm, deformation scale 10. Heat input due to pointwise laser treatment, 25P-pattern geometry. 93

4.61 Load-displacement behaviour of laser treated and untreated plate-structures. Boundary conditions are in each case applied on the undeformed edges (edge-'pullback' denoted by acronym 'PB' in the respective legend). 94

4.62 Load-displacement behaviour of laser treated and untreated plate-structures. The out-of-plane displacements in z-direction are in the current case Δ-values, according to Eq. 4.4. Boundary conditions are in each case applied on the undeformed edges (edge-'pullback' denoted by acronym 'PB' in the respective legend). 95

4.63 Comparison of laser treated and untreated plate-structures in their load-displacement behaviour, boundary conditions applied to the deformed edges. 96

List of Figures

4.64 Load-displacement behaviour of the untreated base plate with included imperfection (1st buckling mode, normalised to maximum transversal displacement of 1 mm, of a perfect, untreated plate scaled with factor $\frac{1}{1000}$) in comparison to two cases of 25P-type laser treated structures. Case 1) boundary conditions applied to structure with edges after 'pullback', case 2): boundary conditions applied to the deformed structure. 97

4.65 Fundamental frequency over the edge line-load for plate-structures subjected to different laser treatments, boundary conditions applied to undeformed edges ('pullback'). 99

4.66 Square of the fundamental frequency over the edge line-load for plate-structures subjected to different laser treatments, boundary conditions applied to undeformed edges ('pullback'). 100

4.67 Fundamental frequency over the edge line-load in for plate-structures subjected to different laser treatments, boundary conditions applied on the deformed edges. 101

4.68 Square of the fundamental frequency over the edge line-load for plate-structures subjected to different laser treatments, boundary conditions applied on the deformed edges. 102

4.69 Fringe plot of temperature field in °C during weak laser treatment of plate A, see for example Chapter 4.5.2, on the plate's top-surface. Laser active for $t = 1.8$ s, deformation scale 1.0. 105

4.70 Fringe plot of temperature field in °C during strong laser treatment of plate AS, see for example Chapter 4.5.2, on the plate's top-surface. Laser active for $t = 1.8$ s, deformation scale 1. 106

4.71 Boundary conditions present for investigations on 'stiffened' plates. a) boundary conditions during laser treatment, b) boundary conditions set for buckling- and eigenvalue analysis. 107

4.72 Direction δ of preset displacement during the buckling- analysis. . . . 108

4.73 First buckling mode of plate PA_U (untreated plate without stiffeners). The FE-mesh used is the same as it serves for treated plates. 110

4.74 Mode of fundamental frequency of plate PA_U (untreated plate without stiffeners). 111

4.75 First buckling mode of plate PB_U (untreated plate with stiffeners). . 112

193

List of Figures

4.76 Fundamental vibration mode of plate PB_U (untreated plate with stiffeners). 113

4.77 First buckling mode of plate PA (unstiffened plate subjected to weak laser treatment). 115

4.78 First buckling mode of plate PAS (unstiffened plate subjected to strong laser treatment). 116

4.79 Longitudinal residual stresses σ_{yy} of laser treated plates A (PA) and AS (PAS). 117

4.80 Longitudinal residual stresses σ_{yy} of weakly laser treated plates A (PA) after complete cooling down in different section points (SP). SP1 (-0.375 mm) plate's bottom-side (averting to laser beam), SP2 (-0.1875 mm), SP3 (0 mm) plate's mid fraction, SP4 (0.1875 mm), SP5 (0.375 mm) plate's top-side. 119

4.81 Longitudinal residual stresses σ_{yy} of strongly laser treated plates AS (PAS) after complete cool-down in different section points (SP). SP1 (-0.375 mm) plate's bottom-side (averting to laser beam), SP2 (-0.1875 mm), SP3 (0 mm) plate's mid fraction, SP4 (0.1875 mm), SP5 (0.375 mm) plate's top-side. 120

4.82 First buckling mode of plate AS (strong laser treatment of unstiffened plate, stiffeners are mounted after laser treatment). 121

4.83 Fundamental vibration mode of plate AS (strong laser treatment of unstiffened plate, stiffeners are mounted after laser treatment). 122

4.84 Boundary conditions in case of an edge-'pullback'. 124

4.85 First buckling mode of plate AS_0 (strong laser treatment of unstiffened plate, out-of-plane deformations, in z-direction, on the edges are reset to zero). 125

4.86 Residual displacements for laser treated structures along $y = 0$. Conditions of structures during laser treatment: A (PA)...unstiffened plate subjected to weak laser treatment, AS (PAS)...unstiffened plate subjected to strong laser treatment, B...weak laser treatment of already stiffened plate, BS...strong laser treatment of already stiffened plate. Pullback of translatoric boundary condition z-direction: AS_0...originally unstiffened plate subjected to strong laser treatment. 128

4.87 Introduction of a continuous laser track into a plate-structure, I-geometry. (courtesy of G. Figala, LUT, Montanuniversitaet Leoben) . 129

List of Figures

4.88 Residual deformations of laser treated plate-structures. Left picture shows plate after laser treatment with circle-geometry. Right picture shows plate after 25 point laser treatment. (courtesy of G. Figala, LUT, Montanuniversitaet Leoben) 130

4.89 Testing machine for plate buckling investigations. (courtesy of G. Figala, LUT, Montanuniversitaet Leoben) 131

4.90 Experimentally obtained load-displacement curves for different plate-structures. (experimental data courtesy of G. Figala and B. Buchmayr, Lehrstuhl für Umformtechnik, LUT, Montanuniversitaet Leoben) 132

4.91 Boundary condition influence on the load-displacement behaviour of a plate-structure (simulation vs. experiment). (courtesy of G. Figala and B.Buchmayr, LUT, Montanuniversitaet Leoben) 133

5.1 Commercial vehicle Citroën Typ H (from a time period between 1947 - 1981, depicted model represents a later version) featuring a bodywork with numerous introduced beads. (courtesy of T. Hatzenbichler, LUT, Montanuniversitaet Leoben) 137

5.2 Two basic methods for producing beads in plane, half-finished products. Left picture: Roll forming of beads. Right picture: Deep drawing of beads. 138

5.3 Idea of the stiffening concept behind the bead laying algortihm. Picture a): Buckling modes for beam with simple supports at both ends and beam with sufficiently stiff additional mid-stiffener-spring. Picture b): Diagram showing normalised critical load over stiffness of mid-stiffener-spring. 139

5.4 Scheme of the incremental bead design procedure. 140

5.5 Concept of the curvature component extraction to maintain the curvature tensor. 142

5.6 Situation on an element patch surrounding a single node. The principal curvature vector is obtained from integration points in each of the surrounding elements and averaged at the node. 143

5.7 Development of a bead laying path. 144

5.8 Boundary conditions for the two investigated structures (plate-structure, U-shaped profile). 147

List of Figures

5.9 First increment of the bead laying algorithm applied to the plate-structure. Left picture: Buckling mode of the untreated plate-structure. Mid picture: Vectors showing maximum curvature values and its respective directions. Right picture: Bead designed in the first increment, shown together with the resulting new buckling mode. 148

5.10 Maximum curvature values and resulting beaded plate-structures with the arising first buckling mode shapes being the basis for the next design increment. Depicted are results of the bead laying algorithm for increments $n = 2$ to $n = 4$. 149

5.11 Development of the critical load concerning plate buckling after each bead laying increment is applied on the plate-structure. 151

5.12 Development of the fundamental frequency of the plate after each increment of the bead laying design is applied on the plate-structure. 152

5.13 Results of the bead laying algorithm applied to the U-shaped profile. Curvature information in each increment is gained from the first buckling mode. 153

5.14 Progress of the bead laying designed for the U-shaped profile. 154

6.1 Kinematics of the roll forming process, symmetrical half model with semi-circular cross section. Distance $\overline{M_1 M_2}$ keeps constant length during deformation, $\overline{AB}$ edge line during forming step. 164

6.2 Results of a FEM simulation showing step-wise deformation between roll stands. In this model edge buckling occurs after passing through the last roll stand. 165

6.3 Assembly of the FEM model showing the main parts of the modelled roll forming mill to simulate the deformation process (semi-circular cross section of finished product). 166

6.4 Schematic diagram of the 'roll flower' applied in the unstable roll forming process to obtain a semi-circular cross-section profile. 166

6.5 Longitudinal stress σ_{11} along profile's edge (path of coordinate s). . . 168

6.6 Longitudinal strain ε_{11} along profile's edge (path of coordinate s). . . 169

6.7 Longitudinal stress σ_{11} through the shell thickness for a mid-element. 170

6.8 Longitudinal stress σ_{11} through the shell thickness for an edge-element.171

6.9 Assembly of the FEM model showing the roll forming mill to simulate the production of the U-shaped profile. 173

List of Figures

6.10 Schematic diagram of the 'roll flower' applied in the simulation of the production of the U-shaped profile. 174
6.11 U-shaped profile during the roll forming process simulation. 175
6.12 Resulting U-shaped profile from the roll forming process simulation. . 175
6.13 Partitioned plate metal structure representing the basic product for the roll forming of the U-shaped Profile. Longitudinal dashed lines show bending zone of profile's flanges, horizontal dashed lines show cutting pathes. 176
6.14 Fringe Plot of longitudinal residual stresses σ_{11} before the cutting operation and afterwards, respectively (grey elements are deactivated).177
6.15 Distribution of longitudinal residual stress σ_{11} along the path s following the cross section in the mid span. Stresses are shown for profile's top-surface, mid-surface and bottom-surface. 178
6.16 Boundary conditions applied during the linear buckling analysis. . . . 179
6.17 First buckling mode. Left picture: 'perfect' U-shaped profile. Right picture: U-shaped profile containing residual stresses due to roll forming. 180
6.18 Boundary conditions applied during the linear eigenfrequency analysis.180
6.19 Fundamental vibration mode. Left picture: 'perfect' U-shaped profile. Right picture: U-shaped profile containing residual stresses due to roll forming. 181
6.20 Boundary conditions for the collapse behaviour investigation for the U-shaped profile. 182
6.21 Collapse behaviour of U-shaped profiles subjected to load case 1 (a prescribed upwards displacement). Picture a): 'perfect' U-shaped profile. Picture b): roll formed profile containing residual stresses. . . 183
6.22 Collapse behaviour of U-shaped profiles subjected to load case 2 (a prescribed downwards displacement). Picture a): 'perfect' U-shaped profile. Picture b): roll formed profile containing residual stresses. . . 184
6.23 Load displacement curves for both load cases, each for the 'perfect' U-shaped profile as well as for the roll formed imperfect and residual stress containing profile. 185

Bibliography

[1] *ABAQUS 6.7, Dassault Systèmes Simulia Corp., Providence, RI, USA, 2007.*

[2] *ABAQUS 6.9, Dassault Systèmes Simulia Corp., Providence, RI, USA, 2009.*

[3] E. Mazza, Methoden der Strukturanalyse, Eidgenössische Technische Hochschule, Zürich, 2008.

[4] *Medtool, Copyright 2006 - 2011, D.P., TU-Vienna, ILSB.*

[5] T. Daxner and F.G. Rammerstorfer, Unterlagen zur Vorlesung Nichtlineare Finite Elemente-Methoden, TU-Vienna, 2011.

[6] Technische Lieferbedingungen kaltgewalztes und oberflächenveredeltes Stahlband - Normenvergleich - Stand Jänner 2004, voestalpine AG, 4020 Linz, Austria.

[7] The Bead Laying Algorithm. private communication with F.G. Rammerstorfer.

[8] A. Bahadur, B.R. Kumar, A.S. Kumar, G.G. Sarkar, and J.S. Rao. Development and comparison of residual stress measurement on welds by various methods. *Materials Science and Technology*, 20:261 – 269, 2004.

[9] E. Beyer and K. Wissenbach. *Oberflächenbehandlung mit Laserstrahlung*. Springer-Verlag Berlin Heidelberg, 1998.

[10] C. Bilik, D.H. Pahr, and F.G. Rammerstorfer. A bead laying algorithm for enhancing the dynamic and stability behaviour of thin shell structures. In *Book of Abstracts of the 37th Solid Mechanics Conference, P. Kowalczyk (ed.), p. 302*, 2010.

[11] C. Bilik and F.G. Rammerstorfer. Instabilitäten während des Walzens von Profilen und im Einsatz von gewalzten Profilen. In *XXX. Verformungskundliches Kolloquium (Ed. B. Buchmayr), Montanuniversität Leoben, p. 93*, 2011.

Bibliography

[12] C. Bilik, F.G. Rammerstorfer, and T. Daxner. Improving the dynamic and stability behaviour of plates by laser treatment. In *Proceedings of the 6th ICCSM (Eds. I.Smojver, J.Soric), paper 104; Croatian Society of Mechanics, Zagreb*, 2009.

[13] C. Bilik, F.G. Rammerstorfer, G. Figala, and B. Buchmayr. Computational modelling of laser treatment of plates for increased buckling loads and natural frequencies. *Journal of Mechanical Engineering Science*, 2011 in press.

[14] M. Brunet, B. Lay, and P. Pol. Computer aided design of roll-forming of channel sections. *Journal of Materials Processing Technology*, 60(1-4):209 – 214, 1996. Proceedings of the 6th International Conference on Metal Forming.

[15] M. Brunet and S. Ronel. Finite element analysis of roll-forming of thin sheet metal. *Journal of Materials Processing Technology*, 45(1-4):255 – 260, 1994.

[16] D. Chapelle and K.J. Bathe. *The Finite Element Analysis of Shells - Fundamentals*. Springer-Verlag Berlin Heidelberg, 2003.

[17] N. Duggal, M.A. Ahmetoglu, G.L. Kinzel, and T.Altan. Computer aided simulation of cold roll forming - a computer program for simple section profiles. *Journal of Materials Processing Technology*, 59(1-2):41 – 48, 1996. Selected Papers on Metal Forming and Machining.

[18] D.M. Emmrich. *Entwicklung einer FEM-basierten Methode zur Gestaltung von Sicken für biegebeanspruchte Leitstützstrukturen im Konstruktionsprozess*. PhD thesis, Institut für Produktentwicklung Fakultät für Maschinenbau Universität Karlsruhe, 2004.

[19] D. J. Ewins. *Modal testing: Theory, practice and application*. Research Studies Press, 2000.

[20] M. Farzin, M.S. Tehrani, and E. Shameli. Determination of buckling limit of strain in cold roll forming by the finite element analysis. *Journal of Materials Processing Technology*, 125-126:626 – 632, 2002.

[21] G. Figala and B. Buchmayr. 1.Zwischenbericht, Fertigungstechnischer Leichtbau Projekt A3.9. Technical report, Montanuniversität Leoben, Department Product Engineering, Lehrstuhl für Umformtechnik, 2009.

Bibliography

[22] F.D. Fischer, E. Hinteregger, and F.G. Rammerstorfer. Numerische Simulation einer experimentellen Spannungsanalyse. *Material Prüfung*, 32. Jahrgang, 1990.

[23] K. Forstner, G. Figala, T. Hatzenbichler, and B. Buchmayr. Analyse und Diskussion verschiedener Verfahren der Eigenspannungsmessung anhand von laserbehandelten dünnwandigen Blechen. In *Tagungsband XXIX. Verformungskundliches Kolloquium*, 2010.

[24] A. Gehring. *Beurteilung der Eignung von metallischem Band und Blech zum Walzprofilieren*. PhD thesis, Fakultät für Bauingenieur-, Geo- und Umweltwissenschaften, Universität Fridericiana zu Karlsruhe, 2008.

[25] R. Görgl. Grundsatzversuche zum Laserhärten von Dünnblechen, Versuchsbericht. Technical report, Joanneum Research Forschungsgesellschaft mbH, Laserzentrum Leoben, 2008.

[26] G.T. Halmos, editor. *Roll Forming Handbook*. CRC Press Taylor & Francis Group, 2006.

[27] F. Heislitz, H. Livatyali, M.A. Ahmetoglu, G.L. Kinzel, and T. Altan. Simulation of roll forming process with the 3-D FEM code PAM-STAMP. *Journal of Materials Processing Technology*, 59(1-2):59 – 67, 1996.

[28] H. Hertel. *Leichtbau*. Springer Verlag Berlin-Heidelberg-New York, 1980.

[29] R.L. Huston and C.Q. Liu. *Principles of Vibration Analysis with Applications in Automotive Engineering*. SAE International, 2011.

[30] L. Ingvarsson. Cold-forming residual stresses. effect on buckling. In *Recent research and developments in cold formed steel structures Proc. Of the Third International Specialty Conference on Cold-Formed Steel Structures*, number 1, pages 85–119, 1975.

[31] L. Ingvarsson. Rullformning av höghållfasta stål. Technical report, AKV-ORTIC AB, Ludvika, 2000. (in Swedish).

[32] S.H. Jeong, S.H. Lee, G.H. Kim, H.J. Seo, and T.H. Kim. Computer simulation of U-channel for under-rail roll forming using rigid-plastic finite element methods. *Journal of Materials Processing Technology*, 201(1-3):118 – 122, 2008. 10th International Conference on Advances in Materials and Processing Technologies - AMPT 2007.

Bibliography

[33] M. Kiuchi. Analytical study on cold roll forming process. Technical report, Institute of Industrial Science, University of Tokyo, 1972.

[34] M. Kiuchi and T. Koudabashi. Automated design system of optimal roll profiles for cold roll forming. In *Procceding of the the third international conference on rotary metalworking processes, Kyoto Japan*, pages 423–36, 1984.

[35] M. Kiuchi, H.M. Naeini, and K. Shintani. Computer aided design of rolls for reshaping processes from round pipes to channel-type pipes. *Journal of Materials Processing Technology*, 111(1-3):193 – 197, 2001.

[36] S.H. Li, G. Zeng, Y.F. Ma, Y.J. Guo, and X.M. Lai. Residual stresses in roll-formed square hollow sections. *Thin-Walled Structures*, 47(5):505 – 513, 2009.

[37] M. Lindgren. Experimental investigations of the roll load and roll torque when high strength steel is roll formed. *Journal of Materials Processing Technology*, 191(1-3):44 – 47, 2007. Advances in Materials and Processing Technologies, July 30th - August 3rd 2006, Las Vegas, Nevada.

[38] J.H. Luo and H.C. Gea. Optimal bead orientation of 3d shell/plate structures. *Finite Elements in Analysis and Design*, 31:55–71, 1998.

[39] E. Macherauch and V. Hauk. Eigenspannungen, Entstehung-Messung-Bewertung. *Deutsche Gesellschaft für Metallkunde e.V.*, Band 1, 1983.

[40] Marco Amabili. *Nonlinear Vibrations and Stability of Shells and Plates*. Cambridge University Press, 2008.

[41] S. Marsoner. Eigenspannungsanalyse an 16 Blechen im Laserspot. Technical report, MCL Leoben, 2008.

[42] W. Mitter, F.G. Rammerstorfer, O. Gründler, and G. Wiedner. Discrepancies between calculated and measured residual stresses in quenched pure iron cylinder. *Material Science and Technology*, 1:793–797, 1985.

[43] G. Nefussi, L. Proslier, and P. Gilormini. Simulation of the cold-roll forming of circular tubes. *Journal of Materials Processing Technology*, 95(1-3):216 – 221, 1999.

Bibliography

[44] A. Otto, A. Kach, and M. Geiger. Eigenspannungen und Verzug durch Wärmeeinwirkung. *Deutsche Forschungsgesellschaft (DFG)*, Kapitel 6, Wiley-VCH-Verlag GmbH:365–391, 2006.

[45] A. Peiter. *Eigenspannungen 1.Art.* Michael Trilitsch Verlag Düsseldorf, 1966.

[46] F.G. Rammerstorfer. Increase of the first natural frequency and buckling load of plates by optimal fields of initial stresses. *Acta Mechanica*, 27:217–238, 1977.

[47] F.G. Rammerstorfer. *Repetitorium Leichtbau.* R. Oldenbourg Verlag Wien München, 1992.

[48] F.G. Rammerstorfer, D.F. Fischer, H. Steiner, W. Mitter, and G. Schatzmayr. Zur Bestimmung der Eigenspannungen in Bauteilen bei Wärmebehandlung mit Phasenumwandlung. *Berichte eines Symposiums in Bad Nauheim, Deutsche gesellschaft für Metallkunde e.V.*, 1979.

[49] F.G. Rammerstorfer and F.D. Fischer. A method for the experimental determination of residual stresses in axisymmetric composite cylinders. *Journal of Engineering Materials and Technology*, 114:90–96, 1992.

[50] F.G. Rammerstorfer and W. Mitter. Finite-Elemente-Modell zur Bestimmung der Umwandlungsplastizität bei Eigenspannungsberechnungen. *Deutsche Gesellschaft für Metallkunde e.V.*, Band 1:239–250.

[51] F. Richter. Die wichtigsten physikalischen Eigenschaften von 52 Eisenwerkstoffen. *Stahleisen Sonderberichte, Verlag Stahleisen GmbH, Düsseldorf*, Heft 8:1–20, 1973.

[52] M. Riehle and E. Simmchen. *Grundlagen der Werkstofftechnik.* Deutscher Verlag für Grundstoffindustrie DVG, 2000.

[53] H. Schimmöller. Berechnung des numerischen Einflusses von Zerlegeschrittweite und Auswertegleichung auf die Bestimmungsgenauigkeit von experimentell-rechnerischen Eigenspannungsermittlungen. *Z.f.Werkstofftechnik (J. of Materials Technology)*, Nr.6 4.Jahrgang:315–322, 1973.

[54] J. Singer, J. Arbocz, and T. Weller. *Buckling Experiments, Experimental Methods in Buckling of Thin-Walled Structures, Volume 1.* John Wiley & Sons, INC., 1998.

Bibliography

[55] M. Tajdari and M. Farzin. Numerical analysis of cold roll forming of symmetrical open sections. *Journal of Materials Processing Technology*, 125-126:633 – 637, 2002.

[56] M.S. Tehrani, P. Hartley, H.M. Naeini, and H. Khademizadeh. Localised edge buckling in cold roll-forming of symmetric channel section. *Thin-Walled Structures*, 44(2):184 – 196, 2006.

[57] M.S. Tehrani, H.M. Naeini, P. Hartley, and H. Khademizadeh. Localized edge buckling in cold roll-forming of circular tube section. *Journal of Materials Processing Technology*, 177(1-3):617 – 620, 2006. Proceedings of the 11th International Conference on Metal Forming 2006.

[58] TRUMPF Werkzeugmaschinen GmbH, Stanzwerkzeuge, D-70839 Gerlingen. *Firma Trumpf Werkzeuginformation Rollabsetzen und Rollsicken*, März 2008. WZ52DE.DOC.

[59] G. Zeng, X.M. Lai, Z.Q. Yu, and Z.Q. Lin. Numerical simulation and sensitivity analysis of parameters for multistand roll forming of channel section with outer edge. *Journal of Iron and Steel Research, International*, 16(1):32 – 37, 2009.

[60] G. Zeng, S.H. Li, Z.Q. Yu, and X.M. Lai. Optimization design of roll profiles for cold roll forming based on response surface method. *Materials & Design*, 30(6):1930 – 1938, 2009.

i want morebooks!

Buy your books fast and straightforward online - at one of world's fastest growing online book stores! Environmentally sound due to Print-on-Demand technologies.

Buy your books online at
www.get-morebooks.com

Kaufen Sie Ihre Bücher schnell und unkompliziert online – auf einer der am schnellsten wachsenden Buchhandelsplattformen weltweit! Dank Print-On-Demand umwelt- und ressourcenschonend produziert.

Bücher schneller online kaufen
www.morebooks.de

VDM Verlagsservicegesellschaft mbH
Heinrich-Böcking-Str. 6-8   Telefon: +49 681 3720 174   info@vdm-vsg.de
D - 66121 Saarbrücken   Telefax: +49 681 3720 1749   www.vdm-vsg.de

Printed by Books on Demand GmbH, Norderstedt / Germany